HOW PATHOGENIC VIRUSES *WORK*

Lauren Sompayrac, Ph.D.

Retired Professor
Dept. of Molecular, Cellular, and Developmental Biology
University of Colorado
Boulder, Colorado

JONES AND BARTLETT PUBLISHERS
Sudbury, Massachusetts
BOSTON   TORONTO   LONDON   SINGAPORE

World Headquarters
Jones and Bartlett Publishers
40 Tall Pine Drive, Sudbury, MA 01776
978-443-5000
info@jbpub.com
www.jbpub.com

Jones and Bartlett Publishers Canada
2406 Nikanna Road
Mississauga, ON L5C 2W6
CANADA

Jones and Bartlett Publishers International
Barb House, Barb Mews
London W6 7PA
UK

Library of Congress Cataloging-in-Publication Data
Sompayrac, Lauren.
    How pathogenic viruses work / Lauren Sompayrac.
        p. cm.
    ISBN 0-7637-2082-8
        1. Host-virus relationships. 2. Virus diseases—Pathogenesis. I. Title.

    QR482 .S65 2002
    616'.0194—dc21

2001050837

Executive Publisher: Christopher Davis
Editorial Assistant: Karen Zuck
Production Editor: Anne Spencer
Manufacturing Buyer: Therese Bräuer
Text Illustrations: Fran Jarvis/ANCO Art
Feature Illustrations: Jim Adams
Typesetting: Carlisle Publishers Services
Cover Design: Anne Spencer
Printing and Binding: D.B. Hess Company

Printed in the United States of America

06 05 04 03 02   10 9 8 7 6 5 4 3 2 1

The authors, editor, and publisher have made every effort to provide accurate information. However, they are not responsible for errors, omissions, or for any outcomes related to the use of the contents of this book and take no responsibility for the use of the products described. Treatments and side effects described in this book may not be applicable to all patients; likewise, some patients may require a dose or experience a side effect that is not described herein. The reader should confer with his or her own physician regarding specific treatments and side effects. Drugs and medical devices are discussed that may have limited availability controlled by the Food and Drug Administration (FDA) for use only in a research study or clinical trial. The drug information presented has been derived from reference sources, recently published data and pharmaceutical tests. Research, clinical practice, and government regulations often change the accepted standard in this field. When consideration is being given to use of any drug in the clinical setting, the health care provider or reader is responsible for determining FDA status of the drug, reading the package insert, reviewing prescribing information for the most up-to-date recommendations on dose, precautions, and contraindications, and determining the appropriate usage for the product. This is especially important in the case of drugs that are new or seldom used.

About the Cover
We are grateful to Dr. David Hockley for contributing a black and white version of the cover photo, showing HIV–1 entering its target cell.

Photo Credits
p. 29, electron micrograph of measles virus budding from the surface of a cell, courtesy Dr. Shpilke Rozenblatt
p. 36, electron micrograph of rotavirus, © Science Photo Library
p. 41, adenovirus model, courtesy Dr. Richard J. Feldmann
p. 41, electron micrograph of adenovirus, courtesy Drs. Nick Wrigley and Robin Valentine

DEDICATION

I dedicate this book to my sweetheart, my best friend, and my wife: Vicki Sompayrac.

Contents

Acknowledgments

I especially want to thank my good friend, Bob Mehler, who read the entire manuscript, offered excellent advice and suggestions, and helped make writing this book fun. I also wish to thank the following people who offered critical comments on various parts of the manuscript: Charles Bangham, Jim Cook, Andreas Dotzauer, Bin He, Thomas Hope, John Kash, Karla Kirkegaard, Mari Manchester, Jack Routes, Aleem Siddiqui, and Ed Watt. Thanks also to Vicki Sompayrac whose wise suggestions helped make this book more readable, and whose editing was invaluable in preparing the final manuscript.

I would like to express my gratitude to Richard Feldmann, David Hockley, Shpilke Rozenblatt, and Nick Wrigley for contributing some of the figures in this book. Their beautiful pictures truly are worth thousands of my words.

Finally, I would like to thank my editor at Jones and Bartlett, Chris Davis. If you want an editor who will make writing a book fun and easy, Chris is your man.

Lecture 0

How to Use This Book

There are many books which employ viruses as tools to teach molecular and cellular biology. This makes sense. A lot of what we know about these subjects was learned by observing how viruses usurp the biochemical machinery of their host cells. However, these texts tend to treat viruses as bit players, focusing mainly on the cells they inhabit. Consequently, such a book can teach you a lot about cell biology, but it usually won't give you much insight into the "mind of the virus." In *How Pathogenic Viruses Work,* the virus will occupy center stage, because my goal is to give you an overall picture of virus-host interactions <u>from the point of view of the virus</u>.

There are also big, heavy books that seem to contain every possible detail about every virus in the universe. These texts are great for reference, but they give the impression that viruses are incredibly complicated and almost impossible to understand. In fact, viruses are quite simple. They really only know how to solve three problems, and the diseases viral infections cause are the consequences—frequently the unintended

consequences—of the different ways viruses solve these problems.

How Pathogenic Viruses Work is written in the form of "lectures," because I want to talk to you directly, just as if we were together in a classroom. In this book, I will focus on the important concepts, and will leave out as much detail as possible. We will also limit our discussion to viruses that cause diseases in humans. Lord knows, there are plenty of them, and to me (and probably to you), these viruses are really the most interesting ones.

Your professor may use this book as the core text for a course, supplementing these lectures with fascinating facts about his or her favorite viruses. Alternatively, your professor may use this book as a course preview, both to provide you with a global view of how pathogenic viruses work, and to give you "pegs" on which to hang more detail as the course progresses.

But no matter how your professor may choose to use this book, you should keep one thing in mind: I didn't write this book for your professor. This book's for you!

Fathoming the Mind of a Virus

We will begin our quest to "fathom the mind of a virus" with two introductory lectures. In the first, we'll start by speculating briefly on where viruses come from. Then we'll discuss three problems which every virus must solve: how to reproduce within a human cell, how to spread to a new human "host," and how to evade host defenses once it gets there. This will bring us naturally to the hypothesis on which these lectures are based: The pathogenic consequences of a viral infection are determined by the particular ways a given virus has "chosen" to solve these three common problems.

To defend themselves against viral attacks, humans have evolved potent, multi-layered defenses. In Lecture 2, we will analyze these defense strategies to see just what viruses are up against. Clearly, evading host defenses has required a lot of "thought" on the part of those viruses which have become successful human pathogens.

Lecture 1

Viral Origins and Lifestyles

What are Viruses?

Viruses are pieces of RNA or DNA enclosed in a protective coat(s). These simple organisms are parasites which have evolved to reproduce inside, and survive outside, the cells they infect. What makes viruses so amazing is that they can do so much with so little—and that they do it so elegantly. For example, in terms of genetic information, hepatitis B virus is the smallest known human virus (i.e., virus that infects humans) with only four genes. Yet despite its dearth of genetic information, hepatitis B virus is one of the world's most deadly pathogens: Over a million people die each year from hepatitis B-associated liver disease.

Where Did Viruses Come From?

Contrary to what Fox Mulder tells us on "The X Files," viruses are not "gifts" to humans from extraterrestrials. In fact, the viruses which now plague humans almost certainly arose from within the cells that make up plants, humans, birds, and animals. I say, "almost certainly," because nobody really knows for sure how viruses evolved, since there are no fossil-like records that can be consulted. My view is that viruses arose because Mother Nature was in a hurry.

During evolution, mutations occur in the genetic code of an organism, and the "fittest" of the resulting mutants are selected for survival. These changes to the genetic code could have been made only one or a few letters at a time—but then the evolutionary process would have taken a very, very long time. So to speed things up, Mother Nature decided to allow whole "phrases" of genetic information to be transferred from one chromosomal location to another. The use of these

"jumping genes" made it possible for complete functional genetic units to be combined to produce proteins that were multifunctional, and for parts of different genes to be spliced together to create proteins that could perform brand new functions. Indeed, it is estimated that over 30% of all human genes bear traces of such "transposition" events.

This gene shuffling greatly accelerated the evolutionary process, but it also provided a mechanism for creating viruses. As transposition took place, genetic information existed free in the cell, unattached to any chromosome. And since viruses are nothing more than snippets of RNA or DNA enclosed in a protective coat, some of this mobile cellular genetic information could be used by "wannabe" viruses to construct their genomes (defined as the sum total of a virus' genetic information). Of course, any mechanism that resulted in genetic information (either RNA or DNA) being "loose" within a cell could have given viruses their start. However, because viral reproduction requires the functions of multiple viral genes, transposition also made it possible for useful genes to be moved adjacent to each other on a chromosome, so that they could be picked up by a wannabe virus as a group. Based on these considerations, I think it can reasonably be argued that the jumping genes which sped up the evolutionary process provided the raw materials needed for the construction of viruses, and that viruses arose as an unavoidable consequence of rapid gene evolution.

If you ask the average person how many different viruses cause disease in humans, he could probably name fewer than a dozen. In actuality, there are over fifty different viruses that can cause human disease. Because there are so many of them, we can conclude that the evolution of pathogenic (disease-causing) viruses was not a rare event that happened only a few times long ago. In fact, many human viruses probably evolved during the last five to ten thousand years, a mere blink of the eye on the evolutionary scale. And viruses are still evolving today.

Although it is impossible to trace with certainty the exact events that led to the "birth" of any particular virus, it is clear that every virus must solve three basic problems. Indeed, the viruses we know today were selected from a much larger crowd of wannabe viruses. Those viruses that survived are the ones which "learned" to solve all three problems.

Three Problems Every Virus Must Solve

No virus carries with it the machinery (e.g., the ribosomes) required to synthesize proteins, and no virus can generate the energy needed to power the copying (replication) of its genetic information. Because they lack the right stuff for their own reproduction, viruses must "hijack" some of the biosynthetic machinery of the cells they infect, and turn these cells into factories that can make many new copies of the virus. So the first problem every virus must solve is how to reproduce during its "visit" inside a human cell. Not only must a virus' genetic information be replicated, but viral proteins must be produced. Some of these proteins direct this replication process. Other viral proteins are used in assembling the coat(s) that protects the viral genetic information once it leaves the cell. So actually, every virus must have two plans: one for copying its genetic information, and another for producing the messenger RNA (mRNA) that will encode the viral proteins. As you will see, the variety of schemes that viruses use to perform these two operations is just staggering.

Each time a cell divides to make two daughter cells, the DNA of the cell must be copied so that each daughter will receive a complete set of genetic information. Because these cellular "copy machines" are readily available, some viruses (those whose genetic information is in the form of DNA) have chosen to use bits and pieces of the cellular DNA replication machinery to help copy their own genomes. This simple strategy has one problem, however: Most cells in a mature human (e.g., the cells that make up your heart) are no longer actively replicating their DNA and dividing. They are "resting." And when the cellular copy machines of resting cells are not in use, they are usually shut down to conserve energy. So a virus that plans on using the DNA replication machinery of the cell it infects either has to figure out a way to give the infected cell a kick to make it turn the

machinery back on; or the virus must bring with it substitutes for those parts of the cellular copy machine that are not working at the time of viral entry. Other viruses, whose genetic information is in the form of RNA, either bring in their own copy machines (e.g., RNA-dependent RNA polymerases) or have genes which encode the proteins required to assemble these copy machines within the infected cell. By "bringing their own," many RNA viruses are able to replicate their genomes even in resting cells.

Cells also have well-established mechanisms that transport proteins made in the cell out to the cell surface. By "borrowing" elements of the transport machinery of cells they infect, completed virus particles can hitch a ride out of infected cells. Indeed, by capturing and modifying cellular genes, viruses have solved the reproduction problem so successfully that most can take over a human cell and use its biosynthetic machinery to produce thousands of new viruses.

The second problem every human virus must solve is how to spread from one individual to another. After all, even if a piece of genetic information evolves so that it can be replicated many times within a human cell, the wannabe virus will die when that human perishes—unless it can find a way to spread to other humans. As you can imagine, this is not a simple problem for a virus to solve. First, the virus must be physically transported from one host to the next. Viruses have solved this "transmission" problem in ways that take advantage of human behavior as varied as coughing or having sex. And once it reaches its new host, the traveling virus must locate cells in which it can reproduce efficiently. As a rule, viruses are pretty picky about which cells they infect—not just any cell will do. To be appropriate for infection, a cell must have receptors on its surface to which the virus can attach, and the biosynthetic machinery within the cell must be compatible with the replication strategy used by the virus. In the human body, there are about 200 different types of cells (e.g., blood cells, liver cells, or lung cells), and a given virus usually will be able to infect only a few of these many different cell types.

Viruses generally like to infect big organs. For example, the surface area of the respiratory tract is larger than a tennis court, so there are lots of cells in the respiratory tract for a virus to infect. The liver contains about one trillion cells, making this organ an attractive target. By infecting large organs which have many cells, viruses can kill or damage (either directly or indirectly) relatively large numbers of cells without doing serious harm to the human host.

This brings up another important point: When a virus succeeds in spreading to a new host, the "uninvited guest" must learn (evolve) to coexist with that new host, at least long enough to reproduce and spread again. For example, a virus which reproduces so efficiently that it kills its host before it has a chance to infect another human just isn't going to make it. To solve this problem of spreading successfully within the human population, viruses have evolved ingenious ways of establishing an unbroken chain of infection, thereby avoiding extinction.

Even if a virus evolves to reproduce and spread efficiently, that virus will never succeed if it is easily repulsed by the defense mechanisms of its host. So the third problem every virus must solve is how to evade host defenses. Every successful virus has evolved strategies that allow it to elude the host's antiviral defenses—at least long enough for it to reproduce and spread from the infected host to a new recipient. Some mechanisms of viral evasion are rather primitive (yet very effective), while others are amazingly elaborate. It is certain that as humans evolved new defenses, viruses were forced to evolve new evasion strategies, and that viruses and their human hosts are still engaged in an ongoing struggle to gain the upper hand.

It is important to understand that viral evasion of host defenses need not be complete. In fact, if a virus were to completely evade host defenses and reproduce unchecked, the virus would probably kill its host before it could spread to another human. All a virus really has to do is evade these defenses long enough either to spread to another host, or to establish a latent or chronic infection within the original host from which the virus can spread at a later time.

Of course, the solutions to these three problems of reproduction, spread, and evasion must be consistent. For example, it wouldn't make much sense for a virus that reproduces only in liver cells to be spread by coughing. Likewise, it wouldn't do for a virus to evolve to reproduce efficiently in cells of the intestines, yet not evolve a strategy that protects the virus from the acidic conditions present in the stomach, the gateway to the intestines. So it's not good enough for a virus just to solve the three problems of reproduction, spread, and evasion. These solutions must fit together in an overall plan of infection.

Viral Pathogenesis

Most human viruses cause some form of disease in their hosts, although some of these pathological conditions affect only a small subset of infected humans, or are so mild as to be virtually undetectable. When you think about it, it certainly isn't in the best interest of a virus to harm its host, since that would be like biting the hand that feeds it. No, viruses don't intend to make their human hosts ill. Rather, the diseases that viruses cause are the (usually unintended) consequences of the way each virus has chosen to solve the three problems of reproduction, spread, and evasion. What this means is that once you understand how a particular virus solves its three problems, you should be able to predict the pathological consequences of the viral infection.

In some cases, the pathology that results from a viral infection is due to the actions of the virus itself (e.g., killing the cells it infects). In other situations, the host's reaction to the virus (e.g., the host immune response) can play a major role in the pathology. Most of the host's defenses against viruses are not finely focused. Consequently, serious collateral damage to the host can result. As my friend, Jim Cook, points out, the host's "shotgun" approach to defending against a viral infection is somewhat like trying to kill a mosquito with a machete—you may kill that mosquito, but most of the blood on the floor will be yours.

Because host defenses frequently determine the pathological outcome of a viral infection, we will devote the next lecture to an examination of host defense mechanisms. Once you see what viruses are up against, you will better appreciate the elegant strategies they have evolved to thwart host defenses.

Lecture

2

Host Defenses

To protect themselves against viral attacks, humans have evolved potent defenses. Over evolutionary time, each new host defense has been countered by new viral evasion strategies. Indeed, this struggle between host defenses and viral counter-defenses continues even today. In subsequent lectures, we will examine some of the strategies which viruses have evolved to evade host defenses, and to keep them "in the game." However, to fully understand how clever these viruses really are, we must first appreciate the power and diversity of the defenses arrayed against them.

Humans have three types of defenses: physical barriers, the innate immune system, and the adaptive immune system. These defenses are arranged in "layers" so that invaders which penetrate one layer are then dealt with by the defenses located in the layer below. It is this multilayered defense which makes it possible for a human to survive the onslaught of tens of thousands of microbial invaders each day.

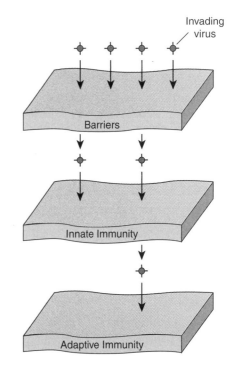

Barrier Defenses

The first defense a virus must penetrate is a physical barrier: the sheets of cells which cover the surface of the body, and which line its internal cavities. The skin is the most difficult of these barriers for a virus to penetrate.

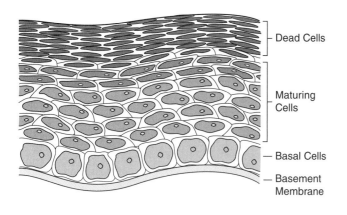

The reason skin is such an effective barrier is that viruses can only infect living cells—and skin is covered with multiple layers of dead cells filled with keratin proteins. In fact, unless skin is punctured or torn, it is essentially impossible for a virus to use this route of infection. As a result, very few viruses have evolved to enter the body via the skin.

Most viruses have chosen to attack the mucosal barrier which lines our respiratory, digestive, and reproductive tracts. These surfaces must be sufficiently permeable that oxygen and nutrients can be transported across them, so some mucosal surfaces are protected by only one or a few layers of living cells. Although an easier target than skin, a mucosal surface still represents a formidable barrier to viral infection.

The Respiratory Tract

Much of the respiratory tract is lined with epithelial cells that have "paddles" (cilia) on their surfaces. These cells are also coated with a blanket of mucus. Indeed, each day about 100 ml of mucus is produced in the airways of a healthy human.

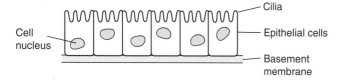

The cilia on these cells have a power stroke that is in the correct direction to move the mucus (and viruses trapped in it) toward the throat, where they can be coughed up or swallowed. Because they must "swim" against this current of sticky mucus, many viruses are unable to get a "grip" on their target cells before they are swept away. Actually "swim" is really the wrong word to use here, because no virus is able to swim. In fact, viruses are totally at the mercy of random motion to bring them close enough to their target cells that they can bind to the appropriate receptor molecules on the cell surface. In addition to the mucosal barrier, some surfaces in the airways are protected by patrolling macrophages, white blood cells which can "eat" (phagocytose) viruses and limit their spread.

The Digestive Tract

Only the toughest of viruses would even consider invading via the digestive tract. To reach the small intestine, where the cells they infect are located, these "enteric" viruses must first survive exposure to saliva which contains compounds that are active against many viruses. So although the mouth seems like it should be an easy target, very few viruses cause infections of the oral cavity because of the potency of antiviral compounds in the saliva. Indeed, recent research suggests that several proteins in saliva (e.g., secretory leukocyte protease inhibitor) are able to inactivate the AIDS virus, and this may explain, at least in part, why HIV-1 infections are not efficiently transmitted during oral sex.

After they battle their way past the saliva, enteric viruses next must brave the acid conditions of the stomach (pH as low as 2.5 is common), and avoid destruction by enzymes like pepsin which thrive at this pH and which are tasked with helping break down proteins in food. Then, as they pass from the stomach into the beginning of the small intestine (the duodenum), viruses must contend with digestive enzymes that are piped into this region from the pancreas. These enzymes, which are designed to disassemble proteins, carbohydrates, and fats, really can do a number on the protein or lipid coats of most viruses. It is also here that bile salts from the liver are added to the digestive mix, and these act as detergents that help break up dietary fats—and viral envelopes.

The Reproductive Tract

Whereas the respiratory and digestive tracts are mainly covered by an epithelial barrier which is only one or a few cells deep, the vagina is protected by an epithelium topped off by multiple layers (a "stratified" epithelium) of relatively flat (squamous), non-proliferating cells.

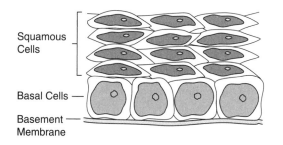

Squamous Cells
Basal Cells
Basement Membrane

Viruses replicate best in human cells that are proliferating, so the multiple layers of non-proliferating cells in the stratified squamous epithelium of the vagina present a rather uninviting target for virus infection. Indeed, viruses which gain entry via the reproductive tract usually rely on small tears in the lining of the vagina (e.g., during sexual intercourse) to allow them access to the proliferating cells in the lower strata of the vaginal epithelium.

The epithelial cells that line the reproductive tract are covered with mucus which helps keep viruses at "arms length." Moreover, the bacteria which colonize the vagina produce lactic acid, and this normally keeps the pH of the vaginal mucus around 5.0. Many viruses are sensitive to acid pH, so this acidic environment is yet another deterrent to viruses which might enter via the reproductive tract.

The Innate Defense System

Of course humans are vastly outnumbered by viruses, so the epithelial barriers cannot be expected to exclude every virus. Indeed, one of the important functions of the physical barriers is to decrease the number of invaders which must be handled by the next line of defense—the innate immune system. This system is called "innate" because it is a defense that all animals seem to have.

The innate immune system includes four main weapons which can be brought to bear, singly or in concert, to defeat invading viruses: professional phagocytes, the complement system of proteins, the interferon system, and natural killer cells.

Professional Phagocytes

Stationed beneath the sheets of epithelial cells are sentinel white blood cells. The most famous of these "professional phagocytes" is the macrophage. These large cells are aptly named because they really are "big

eaters" (phage is from a Greek word meaning "to eat"). Most of the time they patrol the tissues and "collect garbage," including debris from dead or dying cells. However, if they come upon a virus which has made it past the barrier defenses, they will eat it too. Sometimes this is a good thing, and the virus is destroyed by the battery of enzymes inside the macrophage. In other cases, the virus uses the propensity of macrophages to ingest whatever they bump into as an easy way of infecting these professional phagocytes.

Most of the time, macrophages can deal with the occasional virus that slips through the barrier defenses, but sometimes the resident macrophages need reinforcements. Fortunately, the blood is teaming with professional phagocytes which can be summoned by embattled macrophages. These include "young macrophages" (monocytes) and neutrophils—phagocytic cells which exit the blood, ready to kill. About 70% of all the circulating white blood cells are neutrophils—so there is plenty of backup available should the sentinel macrophages get in over their heads.

The Complement System

The complement system is composed of about twenty different proteins that work together to help destroy invaders and to signal other immune system players that the body is under attack. The complement system is very old. Even sea urchins, which evolved about 700 million years ago, have this system. The complement proteins are produced mainly by the liver, and are present at high concentrations in blood and tissues. The most abundant complement protein is called C3, and in the human body, C3 molecules are continuously being broken into two smaller proteins. One of the protein fragments created by this "spontaneous" reaction, C3b, can bind to chemical groups commonly found on the surfaces of viruses.

Once C3b has bound to a virus, this protein fragment can do two important things. First, C3b can catalyze a series of reactions that result in the cutting of many more C3 molecules, creating a "chain reaction" that results in the deposition of large numbers of C3b fragments on the surface of the virus. Because phagocytes (e.g., macrophages) have receptors on their surfaces that can bind to C3b, viruses that have their surfaces decorated (immunologists call it "opsonized") with C3b proteins become preferred targets for phagocyte ingestion. In fact, when macrophages eat a complement-opsonized virus, they are stimulated to become even bigger eaters.

C3b fragments also can bind to the surfaces of "enveloped" viruses whose outer coats are made of membranes picked up from cells they have infected. When this happens, the bound C3b proteins can initiate a biochemical reaction that results in the formation of "membrane attack complexes" which can create holes in the surface of the invading virus. And once a virus has holes in its protective coat, it's all over.

In addition to tagging viruses for digestion and building membrane attack complexes, the complement system has a third important function: Fragments of complement proteins can recruit other immune system players to the battle site. For example, C3a is the piece of C3 that is left over when C3b is made (let nothing be wasted!). Although C3a won't bind to the surface of an invader, it can encourage macrophages and neutrophils to leave the blood and enter the infected area. This "recruiting" function of C3a makes a lot of sense. If many viruses have slipped through the barriers, many molecules of C3 will be cut to produce the C3b fragments needed to opsonize the viruses—and many C3a fragments will be "left over" to summon the phagocytes required to ingest the opsonized virus.

All in all, the complement system is quite versatile. It can destroy enveloped viruses by building membrane attack complexes. It can enhance the function of phagocytic cells by tickling their complement receptors. And it can signal other immune-system cells that the attack is on. Although the complement system and the professional phagocytes are quite effective against viruses which penetrate the physical barriers, these innate system weapons have one major flaw: The complement system and the professional phagocytes can only "get at" viruses when the viruses are <u>outside</u> of human cells. This is a big problem, because a virus-infected cell can produce thousands of new viruses in a period of a few days. And while these viruses are contained within the infected cell, the complement system and the professional phagocytes can't reach them. What is needed are weapons that can destroy infected cells and the viruses which are trying to reproduce within them. Fortunately, the innate system has two weapons which can do just that: interferon and natural killer cells.

The Interferon Warning System

When human cells are under attack by a virus, they can produce "warning proteins" called interferon alpha and interferon beta (the type 1 interferons). Interferon can be made and exported by most cells in the body, and can bind to receptors on the surfaces of nearby, uninfected cells. This binding alerts these cells that they too may soon be attacked by viruses, and that if they are, they must commit suicide. As a result of this altruistic act, the "warned" cells and the viruses that infect them die together, limiting the spread of the virus.

Most, but not all, viruses induce the expression of interferon in the cells they infect. In some cases, the mere contact between the envelope of the virus and the target cell can induce interferon production. More generally, however, it is the presence of a large quantity of double-stranded RNA in virus-infected cells which triggers the cells to produce interferon. Normal cells contain only small amounts of double-stranded RNA. In contrast, many viruses produce huge amounts of double-stranded RNA when they replicate, so double-stranded RNA is a good "clue" that a cell has been infected. In addition, double-stranded viral RNA, produced when viruses attempt to reproduce in cells that have been "warned," can trigger the shutdown of cellular and viral protein synthesis and the destruction of both cellular and viral RNA. Although the shutdown of protein synthesis and destruction of RNA are the two antiviral effects of interferon that have been studied most carefully, there are certainly other ways that interferon "interferes" with viral reproduction.

Natural Killer Cells

The second innate system weapon that can destroy virus-infected cells is the natural killer cell. One of the mysteries about natural killer cells is how they decide which cells to kill. The latest thinking is that this involves two signals—a "kill" signal and a "don't kill" signal—and that it is the balance between these two signals which determines whether a natural killer cell will destroy a potential target. The "kill" signal is thought to involve interactions between proteins on the surface of the natural killer cell and special carbohydrates or proteins on the surface of the target cell. Presumably, these molecules act as flags that indicate the target cell has been infected with a virus, but this part is still poorly understood.

The "don't kill" signal is thought to be the expression of class I MHC molecules on the surface of the potential target cell. These MHC molecules are used by virus-infected cells to present viral proteins, thereby alerting killer T cells—another kind of "killer" cell which we will discuss shortly. Killer T cells are very potent weapons against virus-infected cells, but without class I MHC molecules to present viral proteins, killer T cells are blind to the fact that a cell has been infected. So it makes perfect sense that natural killer cells specialize in killing cells that <u>don't</u> make class I MHC molecules. That way, if a clever virus turns off production of class I MHC molecules in cells it infects, the lack of the MHC "don't kill" signal will make those virus-infected cells susceptible to killing by <u>natural</u> killer cells—so all the bases will be covered.

Cooperation Among Members of the Innate System Team

At least in the test tube, natural killer cells can kill some virus-infected cells without being provoked or "activated," as immunologists like to say. However, natural killer cells kill better if they are activated by signals which let them know there has been a virus attack. Interestingly, one of the most potent activators of natural killer cells are the interferons given off by virus-infected cells. This is an example of the important concept that players on the innate system team cooperate to strengthen the innate defense. For instance, macrophages can be activated to become more voracious "eaters" when complement-tagged viruses bind to their receptors. In addition, activated natural killer cells give off "cytokines"—messenger proteins which cells use to communicate among themselves. These particular cytokines (e.g., interferon gamma—IFN-γ) can help hype up macrophages. And cytokines produced by activated macrophages (e.g., tumor necrosis factor—TNF) can further activate natural killer cells. The result is a positive feedback loop in which activated natural killer cells and macrophages cooperate to get each other fired up.

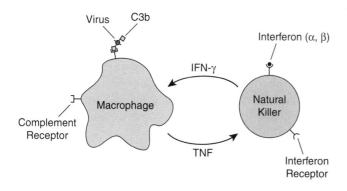

If the number of viruses which get through the barrier defense is small, the invading viruses will simply be opsonized by the complement system and eaten by macrophages. After all, there's no use getting the whole innate system riled up over a few viruses. However, if the attackers are more numerous, cells will be infected, and more viruses will be produced. Many viruses (the "cytolytic" viruses) kill the cells they infect after they have used them as virus factories. This "excess cell death" can alert macrophages that something serious is happening. In addition, the appearance of significant amounts of virus-induced interferon is a warning to natural killer cells that there is a real danger, and that they had better get involved. It is at this point that the innate system players begin to cooperate: More phagocytes are recruited from the blood by complement fragments, and

positive feedback between macrophages and natural killer cells elevates their killing power.

The result of all this cooperation is an "inflammatory response" in which virus-infected cells are killed, and inflammatory cytokines such as tumor necrosis factor and IL-1 are released into the tissues. These inflammatory cytokines can have helpful, indirect effects, causing fever, fatigue, and general malaise. Because some viruses replicate poorly at temperatures above normal body temperature, fever can slow the rate of viral replication. And fatigue and malaise can encourage an infected person to rest and recover at home where he is less likely to spread the viral infection. On the other hand, the inflammatory response can also cause tissue damage because of the local action of inflammatory cytokines and other inflammatory mediators given off by the battling innate system.

The Adaptive Immune System

In many cases, by working together, the innate system players are able to deal with viral attacks. Sometimes, however, the innate system cannot contain the viral infection. This can happen, for example, when the number of viruses present in the initial inoculum is so great that the viruses, replicating to huge numbers in infected cells, simply overrun the innate system. It is at moments like this that the awesome weapons of the adaptive immune system are needed.

B Cells and Antibodies

The adaptive immune system has two main weapons: antibodies and killer T cells. We will begin by discussing antibodies and the B cells that produce them. B cells are "antibody factories" which make antibodies "on demand." Here's how this works.

Like all the other blood cells, B cells are born in the bone marrow, where they are descended from self-renewing stem cells. About one billion B cells are produced each day during the entire life of a human. When a B cell leaves the bone marrow, it has protein molecules called B cell receptors on its surface. Each B cell receptor is constructed from two different proteins, the "heavy chain" and the "light chain," and each of these proteins is made by modular design. For example, in every B cell, on the chromosomes that encode the antibody heavy chain, there are multiple copies of four types of DNA modules (gene segments) called V, D, J, and C. Each copy of a given module is slightly different from the other copies of that module: There are about 100 different V segments, at least four different D segments, six different

J segments, etc. To make the heavy chain gene, each B cell chooses one of each kind of gene segment and pastes them together like this:

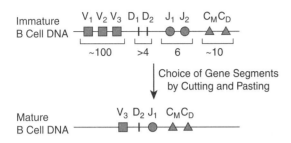

The mature gene for the light chain of the antibody is also assembled by picking gene segments and pasting them together. Because there are so many different gene segments that can be mixed and matched, this scheme can be used to create genes for about 10 million different B cell receptors. And to make things even more diverse, when the gene segments are joined together, additional DNA bases are added or deleted. When this junctional diversity is included, there is no problem creating more than 100 million B cells, each with genes for a different B cell receptor. Indeed, the receptors on different B cells are so diverse that collectively they probably can recognize any organic molecule that could exist. When you consider how many molecules that might be, the fact that a simple mix and match scheme can create such diversity is truly breathtaking.

B cells make "antibodies on demand" by a process known as clonal selection. If the receptor molecules on the surface of a B cell happen to encounter a molecule (called the "antigen") to which they can bind (e.g., a protein on the surface of an invading virus), and if that B cell receives the required "co-stimulatory" signals (more on this later), the B cell will be "selected" to proliferate. This period of proliferation continues for about a week, and results in a "clone" of B cells with identical receptors that can bind to the same antigen. At this time, some cells in this clone begin to produce a form of the B cell receptor which is released into the tissues surrounding the B cell. This secreted form of the B cell receptor is the antibody molecule—a molecule which is identical to the B cell receptor except that it lacks the protein sequences at the tip of the heavy chain molecule which anchor the B cell receptor to the cell surface. A B cell which is committed to antibody production (a "plasma" B cell) can crank out thousands of antibody molecules per second, and this clonal selection scheme insures that only those B cells whose receptors recognize the invader are "selected" to make antibodies.

Antibodies come in five "flavors" (classes): IgM, IgG, IgA, IgD, and IgE. Each of these classes has special properties, but only the first three classes are important in a viral infection.

IgM Antibodies

IgM can be thought of as the "default" antibody class, because IgM antibodies are the ones which B cells first produce after they have proliferated to build up their numbers. These are massive antibodies which have ten identical "hands" (antigen binding regions) that can bind to antigens:

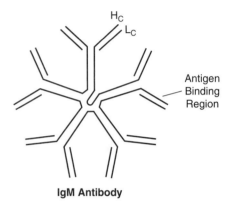

IgM Antibody

IgM antibodies are an excellent choice for the "first antibody" during a viral infection. This antibody class is unparalleled in its ability to activate the complement system and to promote the attachment of complement protein fragments to the surfaces of viruses. Whereas some complement fragments will "spontaneously" attach to the surface of most viruses, when IgM antibodies have bound to these viruses, the complement "tags" are attached much more rapidly. Moreover, some viruses have learned to defend themselves against the spontaneous attachment of complement fragments on their surfaces—and the antibody-mediated deposition of complement fragments can help outsmart these clever viruses. IgM antibodies also can "neutralize" viruses by binding to them and either preventing them from entering their target cells, or by disrupting viral reproduction once the virus has entered.

IgG Antibodies

Although at the beginning of a viral infection, B cells produce IgM antibodies, these same cells can switch to produce other classes of antibodies. This "class switching" usually takes place somewhat later in a viral infection, and results in the production of an antibody class which is even better suited to defend against a particular kind of virus. For example, if a virus enters the blood stream and infects the liver (as hepatitis viruses do), B cells will switch to production of IgG antibodies.

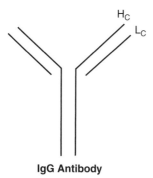

IgG Antibody

IgG antibodies are especially good at defending against pathogens in the blood because they not only can activate the complement system (although not as well as IgM antibodies), but they can efficiently opsonize viruses for ingestion by macrophages and neutrophils. This is because professional phagocytes have surface receptors for IgG, but not for IgM antibodies.

Viruses commandeer the protein-making machinery of the cells they infect, and use it to produce viral proteins. Some viruses then dispatch a subset of these proteins to the surface of the infected cell. In such cases, IgG antibodies that recognize these viral proteins can form bridges between virus-infected cells and natural killer cells (something IgM antibodies cannot do), targeting the virus-infected cells for destruction.

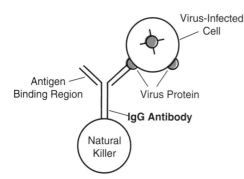

IgG antibodies are also unique in that they can pass from the mother's blood into the blood of the fetus by way of the placenta. This provides the fetus with a supply of IgG antibodies to tide it over until it begins to produce its own—several months after birth. This "passive" immunity can protect the newborn against a variety of viral infections to which the mother has become immune because of previous exposures.

IgA Antibodies

If viruses invade parts of the body protected by mucosal surfaces, B cells will usually switch antibody production from IgM to IgA antibodies—because IgA antibodies are just the ticket for defending against mucosal invaders. One reason for this is that each IgA molecule is like two IgG molecules held together by a "clip."

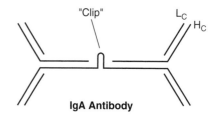

IgA Antibody

This clipped-together tail structure is responsible for several important properties of IgA antibodies. Due to their unique tails, IgA antibodies can be transported from the blood across the intestinal wall and out into the intestine. In addition, whereas each IgG molecule has two antigen binding regions, an IgA molecule actually has four hands to bind to antigens. Because of this feature, IgA antibodies are very good at collecting viruses together into clumps that are large enough to be swept out of the body with the mucus. Their tail structure also makes IgA antibodies resistant to acids and enzymes found in the digestive tract. Notably, it is the IgA class of antibodies that is found in the milk of nursing mothers. These antibodies coat the baby's intestinal mucosa and provide protection against pathogens that the baby ingests.

Although IgA antibodies have qualities which make them particularly suited for protecting mucosal surfaces, they cannot activate the complement system. In addition, professional phagocytes have only low-affinity receptors for IgA antibodies, and consequently, this class of antibody is not well suited to opsonize viruses for phagocytic ingestion. So as you can see, the different classes of antibodies are "specialists" which are particularly useful during certain phases of a viral infection, and for defending against viruses which enter the body by different routes.

During a viral infection, some B cells "choose" not to become antibody factories, but rather to become "memory" B cells. These are the cells which are set aside to protect us against a subsequent attack by the same virus. Memory B cells usually have undergone class switching, so that their function is appropriate for the invader they "remember." In addition, memory B cells usually have "fine tuned" the binding of their receptors by a process called somatic hypermutation. These tuned B cell receptors can bind to the tiny amounts of virus which typically are present at the beginning of an infection. When binding occurs, the memory B cells proliferate rapidly, and begin to produce huge quantities of antibodies which are exactly right to defend against that particular virus.

So not only is the adaptive immune system capable of producing antibodies that can bind to any virus which might evolve, the system also adapts by producing the class of antibody which is just right for defending against a given virus—be it a respiratory virus, a blood-borne virus, or a virus that attacks the digestive system. Finally, the adaptive immune system remembers previous invasions, and can quickly mobilize the appropriate weapons to defend against a subsequent attack. Clearly a virus has its work cut out for it if it is to evade the antibody defense (sometimes called the "humoral" defense) and establish a successful infection.

Antibodies are very effective at tagging viruses for destruction and for neutralizing their infectivity. It is for this reason that antibodies usually play the major role in protecting us against subsequent infections by the same virus. Indeed, it is the goal of most vaccination strategies to produce protective antibodies. However, during an initial viral infection, B cells must be selected, and then proliferate to build up their numbers before antibodies can be produced. This process generally takes a week or more. So by the time antibodies are made, the invading virus may have completed multiple rounds of infection and reproduction. And although the antibodies, when they finally arrive on the scene, are very effective against the newly produced viruses, they have very limited capabilities for facilitating the destruction of virus-infected cells. What is needed is a weapon which can "look inside" cells, and efficiently destroy those which have been infected by a virus. And that weapon is the killer T cell (frequently called a cytotoxic lymphocyte or CTL).

Killer T Cells

Class I MHC molecules are "billboards" that display, on the surface of a cell, a "sampling" of all the proteins that are being manufactured inside that cell. This includes ordinary cellular proteins like enzymes and structural proteins, as well as proteins encoded by viruses and other parasites that might have infected the cell. For example, during an infection, viral proteins are produced using the biosynthetic machinery of the virus-infected cell. Samples of these viral proteins are then chopped up by cellular enzymes, and the resulting protein fragments are loaded onto empty class I MHC molecules. Once loaded, the class I MHC

molecules proceed to the cell surface where they display the protein fragments they are carrying.

Killer T cells are white blood cells that are made in the bone marrow, and which mature in the thymus. Like B cells, killer T cells also have receptors on their surfaces which are made by mixing and matching gene segments. As a result, T cell receptors are about as diverse as B cell receptors. T cells also obey the principle of clonal selection: If a T cell's receptors bind to protein fragments displayed by class I MHC molecules, and if that T cell also receives the appropriate co-stimulation (more on this part later), the T cell will be "activated" and will proliferate to build up a clone of identical T cells. This proliferation stage takes a week or more to complete, so like the antibody response, the T cell response is slow and specific.

Once they have proliferated, killer T cells use their T cell receptors to "scan" the protein fragments displayed by the class I MHC billboards to determine whether a cell has been infected by a virus and should be killed. Almost every cell in the human body expresses class I MHC molecules, so most cells are an "open book" that can be checked by killer T cells. And when a killer T cell detects a virus-infected cell, it destroys it.

So antibodies and killer T cells provide humans with a two-pronged defense. Viruses which are outside of cells are tagged by antibodies and are dealt with harshly. Cells which have been infected with viruses are destroyed by killer T cells, and the viruses within these cells die with them.

Activating the Adaptive Immune System

The incredible diversity of B and T cell receptors is key to having a system that can adapt to defend against any virus. However, this same diversity creates a potentially dangerous situation, since there are also certain to be B and T cells with receptors that will recognize our own "self" molecules. Imagine, for example, how devastating it would be to have B cells that produced antibodies which tagged the insulin proteins in our blood for destruction.

To deal with the possibility of self-reactive B and T cells—cells which potentially could cause autoimmune disease—a screening process is in place which eliminates or incapacitates most B and T cells that could recognize our self antigens. However, the procedures for "teaching" B and T cells self-tolerance are not foolproof, and potentially self-reactive B and T cells still circulate in the blood of every normal human. To reduce the possibility that these cells will cause autoimmune disease, several additional safeguards are imposed on the system. Certainly the most important of these involves limitations on the conditions under which B and T cells can be activated.

Before B cells can produce antibodies, and before killer T cells can kill, they must be activated. As we discussed, the first step required for activation is recognition of antigen by the B cell's receptors or recognition by the T cell's receptors of antigen presented by MHC molecules. However, antigen recognition is not enough for activation. A second, "co-stimulatory" signal is required. And this co-stimulatory signal is only given when the body is under attack and is in real danger. Here's how this works.

When a virus breaches the barrier defenses, the innate immune system springs into action. Indeed, one of the greatest advantages of this system is that it can respond so quickly. The innate system's two main "first response" weapons, the complement proteins and the sentinel macrophages, take care of small infections, usually before many cells can be infected.

If this first response is not enough, more weapons of the innate system are brought to bear. These include activated natural killer cells and additional professional phagocytes (neutrophils and macrophages) which are recruited from the blood. In a viral infection, there are two main clues the innate system uses to decide whether additional weapons are needed: the production of interferon, and the death of virus-infected cells. If cells are producing a lot of interferon and many are being killed by viruses, it's pretty clear that the first response weapons are not going to be enough.

In most cases, the activated natural killer cells and additional phagocytes are sufficient to quell the invasion. Sometimes, however, the viruses are too numerous (or too devious) for the innate system to handle quickly, and the soldiers of the innate system are forced to battle on. During the protracted struggle, cells of the innate system produce large amounts of "battle cytokines" such as interferon gamma and tumor necrosis factor. These battle cytokines are "danger signals" which can activate the <u>adaptive</u> immune system during a viral attack. The production of large quantities of battle cytokines is a clear indication that the innate system has recognized a danger, but is unable to deal with it.

Residing in the tissues that underlie all the exposed surfaces of the body is a very important type of cell—the dendritic cell. It is this cell which is tasked with communicating to the <u>adaptive</u> immune system the danger signals from the <u>innate</u> system. In normal tissues (tissues that haven't been infected), dendritic cells are wildly phagocytic—they take up about four times their own volume of fluid every hour. Mostly, they just drink it in and spit it back out. If, however, the innate system senses grave danger and begins to pump out battle cytokines, the lifestyle of the dendritic cell changes dramatically. In response to danger signals connected with

the virus attack (e.g., tumor necrosis factor), dendritic cells cease phagocytosis, exit the tissues where the battle is raging, and migrate through the lymphatic system to the nearest lymph node. Lymph nodes are "dating bars" where cells of the adaptive immune system meet to be activated. There the migrating dendritic cell uses its MHC molecules to display viral antigens to T cells, and provides the all-important co-stimulatory signals which are required for activation of the adaptive system. Importantly, dendritic cells only travel to lymph nodes and activate the adaptive immune system if a battle is on. This feature helps insure that the adaptive system, with its powerful weapons, only gets involved when the innate system has "certified" that there is an attack. This certification requirement helps safeguard against autoimmunity: The innate system normally does not view our own antigens as dangerous, and as a result, potentially self-reactive B and killer T cells don't get activated.

Selection of Weapons

The adaptive immune system has a number of different weapons it can bring to bear during a viral attack. So how does it know which weapon to use against a given virus? For example, it wouldn't make sense for B cells of the adaptive system to produce a lot of IgA antibodies in response to a blood-borne virus. This antibody class is great against mucosal invaders, but isn't very useful against viruses that have entered the body on a contaminated needle. What is needed in that case are IgG antibodies, which are excellent at defending against blood-borne pathogens. It turns out that the innate system not only activates the adaptive system when there is danger, but it also "instructs" the adaptive immune system on which weapons to make and where to send them. Here's how this works.

Because dendritic cells have observed the battle first hand, they not only know that there has been an attack, but they also have "intelligence" about the nature of the invader (e.g., is it a virus or a bacterium) and where the invader entered the body (e.g., via the respiratory system or through the blood). In addition, some viruses can infect dendritic cells before they leave the battle site, allowing dendritic cells to pre-sent the adaptive immune system with a clear picture of what an infected cell "looks like." Once all this intelligence on the invader has been gathered, dendritic cells produce a "battle plan." Because of this ability to collect data and synthesize a plan of action, dendritic cells are considered to be the most important of all the "antigen presenting cells."

In the lymph nodes, the defense strategy devised by the dendritic cell is communicated to another player on the adaptive immune system team, the helper T cell. It is the task of the helper T cell to implement the plan and to mobilize the weapons (e.g., antibodies or killer T cells) that are appropriate in that particular instance. The details of how this defense strategy is communicated to helper T cells, and how they in turn elicit the appropriate immune response are not well understood. It is known, however, that the communications between dendritic cells, T cells, and B cells involve either direct contact between these cells or the exchange of cytokines, or both.

So the dendritic cell can be thought of as the "coach" of the immune system team. It collects information about the opponents and formulates a "game plan." This plan is then given to the helper T cell, the "quarterback" of the team, who calls the signals and sends the proper players "into the game." This insures that the response of the adaptive immune system is appropriate to the invader—be it a virus that enters via the respiratory tract, the digestive tract, the reproductive tract, or the blood.

Weaknesses of the Adaptive Immune System

In terms of defending against a viral infection, the adaptive immune system, although incredibly powerful, has at least three weaknesses. First, the adaptive immune system is slow to react. Humans have evolved a multi-layered defense against viral attacks, and the adaptive immune system represents the third layer. So when a virus attacks for the first time, the adaptive immune system usually gets involved only when the innate system is being overwhelmed. Furthermore, because the weapons of the adaptive system—antibodies and killer T cells—must be custom-made in response to each particular virus, the adaptive immune system takes a week or two to mount a strong defense against a virus it has never seen before. In contrast to the slow reaction of the adaptive immune system, viruses work quickly. Unchecked, a single infecting virus could easily produce billions of new viruses before the adaptive immune system could reach full strength. It is for this reason that the innate system, which can spring into action immediately, is so important in the initial phases of a virus infection.

The innate system is very good at recognizing danger from invading bacteria. Many of these bacteria have unusual carbohydrates on their surfaces for which immune system cells (e.g., macrophages) have receptors. When these receptors bind to the bacterial carbohydrates, the innate system immediately springs into action. In contrast, most viruses do not have "danger signatures" which can be recognized by receptors on immune system cells. As a result, the innate system usually becomes aware of a viral infection only when virus-infected cells give off interferon, or when viruses kill the cells they infect. What this means is that viruses which do not induce interferon production and which do not kill their target cells can sneak in "under the radar" of the innate system. And if the innate system doesn't sense danger, the adaptive system doesn't get activated. So the lack of a "danger signature" which can be directly recognized by the innate system is a weakness that viruses can exploit.

Finally, the adaptive immune system functions less well in infants and in aged humans. In fact, the adaptive immune system reaches its peak power at about puberty, and then it's downhill from there. As we get older, our thymus functions less and less efficiently. Since this is the organ which turns out new T cells, this means that older humans are less and less able to respond to viruses which they have never encountered before. However, before we complain that the system was poorly "designed," we must remember that these immune defenses evolved to protect humans only until they were old enough to reproduce and raise their offspring. Consequently, the health of old people was really not a consideration.

Although our antiviral defenses have a few flaws, they are wonderfully effective at preventing most viruses from enjoying a lengthy stay in their human hosts. In rare cases, viruses establish "latent" infections in which the virus "hides out" somewhere in the body, or "chronic" infections in which the virus carries on a protracted conflict with the innate and adaptive immune systems. But for the most part, viral infections are handled by host defenses in just a few days or weeks. Actually, from the point of view of the virus, this isn't all that bad. Most viruses need only survive the host defenses long enough to reproduce and spread to a new human. They are like good guests who try not to overstay their welcome.

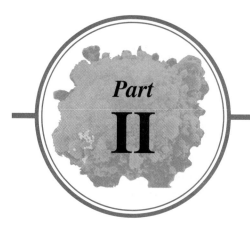

The Bug Parade

Now we come to what my friend, Tom Hill, calls the "Bug Parade"—the part where we talk about individual viruses. Although it depends a bit on how you count them, there are at least fifty different viruses that cause diseases in humans. Any good, thick, book on virology will list all these viruses and will tabulate their main attributes. So if you need to find out everything that is known about a particular virus, you'll want to consult one of those books.

On the other hand, if your intention is just to understand how viruses work, you don't need to know every possible thing about every virus. In fact, I maintain that if you examine the most important features of a dozen, well-chosen viruses, you will be well on your way to understanding the "mind of the virus." And once you understand how viruses "think," all that data in the thick textbooks will make a lot more sense. So here, we will focus on only twelve different viruses and the strategies they use to solve the three problems that all viruses face: reproduction, spread, and evasion. Moreover, I won't try to tell you everything about these twelve viruses. That would be boring. These viruses are examples, after all, so I'll pick and choose what I think are the most interesting aspects of each virus—paying special attention to those features of virus lifestyles that have the greatest impact on determining the diseases these viruses cause.

Most of the time viruses are discussed according to their family classification. In this book, however, I'm going to parade the viruses according to where they first enter the body. The reason for this is that the route of entry strongly influences how a virus reproduces, how it spreads, and what defenses it must evade. Although viruses sometimes are versatile enough to use several ports of entry, I will position each virus in the Parade according to what I consider its most "natural" route of infection. For example, a particular virus might be spread either during the birth of a child (a very natural route, you must agree) or by the sharing of contaminated needles. Such a virus would be grouped with "Viruses We Get From Mom," since clearly viruses did not evolve to use drug abuse as their natural mode of transmission.

After we discuss our twelve model viruses, we will also take a look at some of the "emerging" viruses. Although most of these viruses have not been studied intensively, they can cause serious disease, and are likely to be much more widespread in the future—so it's important that we try to understand them too.

As the twelve model viruses pass in review, I'm sure you'll join me in marveling at the elegant strategies they have evolved to solve their problems. And I hope you will also come to appreciate that it is the nature of these solutions which actually determines the pathological consequences of a viral infection. Let the Parade begin!

Viruses We Inhale

The first bugs to pass in review are respiratory viruses—viruses we inhale. It's natural that these viruses should be positioned at the head of the Parade, because humans acquire more viral infections via this route than by any other. In addition to the obvious cold and flu viruses, respiratory viruses also cause most of the common childhood viral infections—mumps, measles, chicken pox, German measles, and the like.

One reason for the popularity of the respiratory route of infection is simply that respiratory infections are hard to avoid. Good hygiene can greatly lessen the probability of infection by most viruses that enter via the digestive tract. A monogamous lifestyle can virtually eliminate sexually transmitted viral diseases. Modern living conditions can decrease the incidence of viruses transmitted by mosquitos, ticks, or fleas. But if you enter a room in which someone who has a respiratory infection has just coughed or sneezed, just about all you can do is hold your breath. And as more and more people live closer and closer together, respiratory infections become ever more difficult to avoid.

Another reason that so many viruses choose to enter via the airways is that it really is "the easy way in." Although much of the airways is coated with mucus, respiratory viruses can overpower this barrier defense because the droplets generated by a cough or sneeze contain a very large number of virus particles—so many that at least some of these viruses usually will be able to penetrate the mucus and infect the underlying epithelial cells. In contrast, relatively few viruses are able to withstand the acidic conditions in the stomach through which they must pass to infect cells in the intestines. Likewise, conditions in the mouth are inhospitable to most viruses, and unless skin is punctured or abraded, it represents a barrier which no virus can penetrate.

In this lecture we will discuss three viruses we inhale: influenza virus, rhinovirus, and measles virus. These three offer excellent examples of the diverse strategies employed by respiratory viruses to solve their problems of reproduction, spread, and evasion.

INFLUENZA—A BAIT AND SWITCH VIRUS

Influenza is a respiratory virus which is a showcase for viral cleverness. Actually there are three types of human influenza viruses: A, B, and C. Influenza A is the most dangerous of the three, so we will use it as our "prototype" influenza virus. However, influenza B and C viruses also have features that make them interesting to discuss, especially when their properties contrast with those of influenza A. All three types can cause typical flu symptoms, yet these viruses are sufficiently different that antibodies against one type will not protect against the others.

Viral Reproduction

Because viruses can only reproduce using a cell's biosynthetic machinery, all viruses must somehow get their genetic information across the cell's plasma membrane. The first step in this process always involves binding of the virus to receptor molecules on the cell surface. For influenza to infect a cell, a protein on the surface of the virus called the viral hemagglutinin must "plug into" a receptor on the surface of the target cell. Although it is known that the "socket" for the hemagglutinin protein is a sialic acid (a.k.a. neuraminic acid) residue, the cell surface molecules to which these sialic acids are attached (either a glycoprotein or a glycolipid) have not been identified. Interestingly, the hemagglutinin protein got its name when it was discovered that influenza virus could cause red blood cells to clump (agglutinate). This happens because hemagglutinin proteins on the surface of the virus (there are about 500 such proteins per virus) bind to sialic acid residues on the surfaces of red blood cells, forming "bridges" that can connect many red blood cells together.

Once a virus binds to its target cell, it is faced with the task of entering the cell and removing its protective coat so that its genetic material can be replicated. What makes this process so interesting is that the viral coat must be stable enough to protect the viral genome from the harsh conditions that exist outside cells; yet once the virus has reached its target cell, the coat must be at least partially disassembled to allow replication of the viral genome. Viruses have two general strategies for entry and uncoating. In the first, the virus binds to the cell, "checks its coat at the door," and injects its genome into the cytoplasm of its target cell. We will discuss several viruses that use this approach.

The second strategy for entry and uncoating, called receptor-mediated endocytosis, is employed by influenza virus. During this process, the virus binds to receptors on the cell surface, and then is taken into the cell (endocytosed), completely enclosed in a portion of the cell's plasma membrane (called an endosome).

normally transport protons out of the cell. However, when the endosome is formed, the membrane is inverted, so that protons are now pumped into the interior of the endosome. As a result, the environment within the endosome becomes progressively more acidic. And when the pH inside of the endosome reaches about 5.0, a conformational change takes place in the virus coat which allows the coat and the endosome to fuse, releasing the viral genome into the cytoplasm.

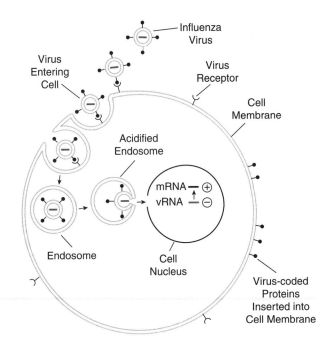

So during receptor-mediated endocytosis, it is the "acid bath" that influenza virus takes which converts its coat from an extremely stable structure to a relatively destabilized configuration from which the viral genetic information can be released.

The genome of influenza virus is comprised of short pieces of single-stranded RNA (e.g., type A has eight such segments). Once released into the cytoplasm, these RNA segments rapidly enter the nucleus of the cell, being directed there by nuclear localization signals present on viral proteins that remain bound to the segments of RNA. It is in the nucleus of the infected cell that replication of the viral genome takes place.

The genomes of single-stranded RNA viruses can be either "positive strand" or "negative strand." Positive-strand viral RNA is defined as RNA which is ready to be translated into proteins—so it is synonymous with viral messenger RNA. Negative-strand RNA is the "opposite" strand, so a complementary copy of it must be made to obtain viral messenger RNA. Influenza is a negative-strand RNA virus.

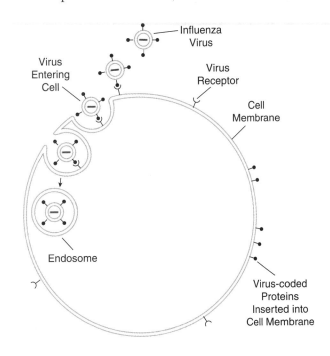

This process of endocytosis gets the virus into the cell, but the flu virus still must figure out a way to shed its protective coat. In addition, it must somehow arrange to get its uncoated genome out of the endosome. The plasma membrane of a cell contains "pumps" which

Because human cells do not have polymerase enzymes that can replicate RNA molecules, every virus which has an RNA genome must encode its own polymerase. Positive-strand RNA viruses can use the cell's protein-making machinery to translate their viral RNAs into proteins, making it possible to produce the required viral polymerase proteins <u>after</u> infection. In contrast, the RNA of negative-strand viruses cannot be translated into proteins, so negative-strand viruses such as influenza must carry their polymerase proteins with them inside their protective coats. Indeed, during an influenza infection, a polymerase molecule remains associated with each influenza RNA segment as it enters the cell nucleus. Once inside, the viral polymerases spring into action and make complementary copies of the viral RNAs to yield positive, protein-coding, messenger RNA strands.

During synthesis of this messenger RNA (mRNA), influenza virus does something rather nasty: Viral proteins actually bite off one end of <u>cellular</u> mRNA molecules that are present in the cell's nucleus, and use these snippets of RNA to begin synthesis of viral mRNAs.

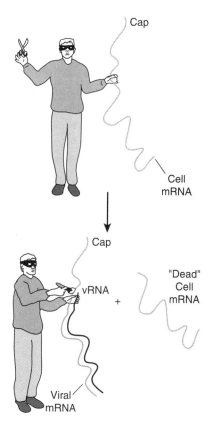

The pieces of cellular messenger RNA that are "stolen" by the virus contain the "cap" structure that ribosomes need to initiate protein synthesis. Because cellular mRNAs whose caps have been stolen can't be translated into protein, cap stealing not only provides the virus with a ready-made cap "for free," it also helps focus the protein synthesis machinery of the cell on the production of viral proteins. Due to cap stealing and other dirty tricks, the takeover of infected cells by influenza virus is so devastating that these cells die as newly made viruses are released. Viruses such as influenza, which kill the cells they infect, are called "cytolytic" viruses.

Some of the viral messenger RNAs are translated in the cytoplasm to make new viral polymerase molecules, which enter the nucleus and produce full-length complementary copies (cRNAs) of the original viral RNAs. These positive-strand cRNAs are then recopied many times by the viral polymerase to make the negative strands that will form the genomes of newly minted viruses. These new viral RNA segments (vRNAs) are then coated with virus-encoded proteins, exit the nucleus, and proceed to the inner surface of the plasma membrane. There, after the requisite number of gene segments has been gathered together, the RNA-protein complex buds from the cell surface, picking up a patch of cell membrane which forms a viral "envelope." Prior to budding, several viral proteins (including the viral hemagglutinin) are inserted into the cell membrane, so the viral envelope actually contains viral proteins in addition to the normal cell membrane components. Here's a diagram that shows roughly how this works. For clarity, I have shown only one RNA segment and have omitted the viral proteins that associate with the viral RNA.

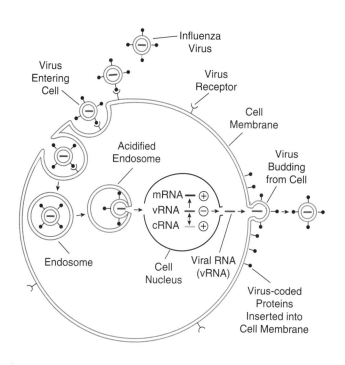

When influenza virus particles get ready to exit an infected cell, they face an "escape" problem. This difficulty arises because the infected cell has receptors on its surface that contain sialic acid molecules to which the hemagglutinin protein of the virus can bind (that's how the virus entered the cell); and newly minted viruses have both hemagglutinin molecules and hemagglutinin receptors on the surface of their envelopes. As a result, exiting viruses can be "captured" by the infected cell when the viral hemagglutinin molecules "re-bind" to the hemagglutinin receptors on the cell surface. In addition, exiting virus particles can clump together when their hemagglutinin molecules bind to receptors on the envelopes of other exiting viruses. To solve this "stickiness" problem, the virus produces a second protein which, like the hemagglutinin protein, is inserted into the cell membrane during infection. This "neuraminidase" protein functions as a "razor" that enzymatically shaves off sialic acid residues from the surface of an infected cell. Once these sialic acid molecules have been removed, viruses can bud through the "shaved" membranes and escape, unencumbered by the sticky receptors.

Because it carries its own RNA polymerase, influenza virus can efficiently infect human cells that are not proliferating. However, the viral polymerase does not have the capacity to "proofread" its work, and mutations arise as incorrect nucleotides are inserted into the growing RNA chains. In fact, the influenza RNA polymerase makes roughly one mistake in every 10,000 nucleotides it copies. Because the flu genome is made up of about this many nucleotides, nearly every virus produced in an influenza-infected cell will be a mutant.

Viral Spread

In humans, influenza virus is usually spread when microdroplets containing virus particles are created by a sneeze or cough of an infected individual—and then are inhaled by another person. Because of their small size, these inhaled microdroplets can sometimes penetrate all the way to the lungs, so influenza can infect epithelial cells that line both the upper and the lower respiratory tract. Once a cell has been infected, it takes only about five hours for thousands of new viruses to be produced. These newly minted viruses go on to infect other cells of the respiratory tract. And because sialic acid molecules are present on all airway epithelial cells, essentially the whole respiratory tract is fair game for an influenza virus infection. In fact, an influenza infection frequently spreads until almost every cell in a branch of the airways has been infected.

Although influenza virus is usually transmitted directly by coughing, the virus can survive for several days on a nonporous surface (e.g., a telephone or a doorknob), so it also can be spread by less direct routes. Fortunately, the oily viral envelope is sensitive to detergents, so hands and other surfaces can be easily disinfected.

When cells of the respiratory tract are killed by the virus (or by the immune response to the virus), the resulting inflammation triggers the cough reflex that is designed to clear the airways of foreign invaders. As a result of this cough, microdroplets containing virus particles are expelled into the surroundings. So as a natural consequence of infecting and killing cells in the respiratory tract, the virus triggers the cough reflex which facilitates its spread. When you consider that a rather heroic cough can expel virus-containing droplets at about 100 miles per hour, that each droplet can contain millions of viruses, and that as few as ten viruses can initiate an influenza infection, it's easy to understand how efficient this mechanism of spread really is. In fact, in settings in which people are living close together (e.g., in a nursing home), upwards of 80% of the potential recipients can be infected during a flu outbreak.

Strains of type A influenza infect animals as diverse as birds, pigs, seals, horses, and ferrets. Type B and type C influenza viruses are primarily human viruses, although they occasionally have been isolated from animals, but not from birds. Different influenza virus strains can have slightly different hemagglutinin molecules, and as a result, the host range (i.e., the type of cell or the species of host it infects) of a particular virus strain can be limited by the ability of the viral hemagglutinin to fit with the different types of sialic acid-containing molecules available on the surfaces of host cells.

In addition to the requirement that its target cell must have a receptor to which the virus can attach, the host range of an influenza virus strain is also restricted by the compatibility of virus-encoded proteins with the biosynthetic machinery of the host cell. For example, for the virus to succeed in entering a cell, the hemagglutinin protein must first be cut into two pieces. This cleavage can be carried out either by enzymes that are inside the cell which produces the virus, or later by enzymes that are in the environment outside the cell. Only certain kinds of cells produce enzymes that can do this cutting, and only certain cellular environments contain the needed enzymes. As a result, the requirement for hemagglutinin cleavage can limit the types of cells an influenza virus can infect. Interestingly, most human cells, including airways epithelial cells, lack the enzyme (a serine protease) required to cleave the hemagglutinin proteins of the strains of type A influenza that normally infect humans. This means that the all-important cleavage step must be performed after newly made viruses have been released into the respiratory tract. Fortunately, there are cells in the airways (e.g., Clara cells) that produce serine proteases and export them out into the mucus. So in the respiratory tract, influenza virus has no problem finding the required enzymes.

Evading Host Defenses

Although the genome of influenza virus is comprised of single-stranded RNA, during replication, when viral RNA is used as a template to make positive-strand RNA, stretches of double-stranded RNA exist in the nucleus of influenza-infected cells. Because of the double-stranded RNA present during viral replication and influenza's lipid envelope, influenza virus is a potent inducer of interferon production. In fact, influenza-infected cells produce such large quantities of interferon that you might expect that the ability of influenza virus to infect neighboring cells would be severely limited by the interferon warning system. However, influenza has evolved mechanisms that can at least partially disable the interferon system in influenza-infected cells. Here's how this works.

When cells are "warned" by interferon binding to their receptors, they begin to produce relatively large amounts of a protein kinase called PKR. This remarkable protein has the property that if it binds to double-stranded RNA (e.g., replicating influenza RNA), it is activated, and in its activated form, it can attach phosphate molecules to certain other proteins. So PKR is induced in interferon-warned cells, and then functions as a sensor that detects double-stranded RNA.

One of the targets that PKR phosphorylates is a protein (eIF2) which is required to initiate protein synthesis. And when eIF2 has those extra phosphates attached, it can no longer participate in the production of proteins. So when a virus infects an interferon-warned cell and produces double-stranded RNA, PKR is activated, eIF2 is phosphorylated, and protein synthesis is shut down. Because both the infected cell and the virus need protein synthesis to survive, this interferon defense leads to the death of the infected cell and the virus trying to reproduce within it.

Influenza virus counters the interferon-induced shutdown of protein synthesis by producing a protein (NS1) that binds to double-stranded RNA, and keeps it from activating PKR. Although influenza's efforts to evade the interferon system are not completely effective, they do buy enough time for the virus to complete its replication cycle within the infected cell. And that's all that really matters.

Because flu virus kills the cells it infects, it doesn't take the immune system long to discover that there has been an invasion, and to take steps to destroy the virus. In fact, the immune response is usually so vigorous that most flu viruses within the body are eliminated within about two weeks. When a person is first exposed to influenza virus, the most important player on the adaptive immune system team is the killer T cell, which can kill virus-infected cells and the viruses reproducing within them. However, during a subsequent exposure to the same virus strain (e.g., during the next flu season), memory B cells and the neutralizing antibodies they produce are the most important defenses against infection. Interestingly, although neutralizing antibodies bind to influenza virus particles, they do not prevent the virus from entering its target cells—a feat that would be extremely difficult considering that each virus envelope displays hundreds of copies of the hemagglutinin "docking" protein. Rather, neutralizing antibodies bind to the hemagglutinin molecules before the virus enters the cell, hang on during entry, and somehow interfere with viral reproduction.

The viral neuraminidase protein is also a target for neutralizing antibodies. Although anti-neuraminidase antibodies can't block infection, they can bind to the neuraminidase "razor" and prevent it from functioning efficiently, thereby severely limiting the number of virus particles that escape from infected cells. Because anti-hemagglutinin and anti-neuraminidase neutralizing antibodies both help protect against an influenza attack, flu strains are defined by the neutralizing antibodies that bind to each of these proteins. For example, the strain of type A influenza called H3N2 can be neutralized by the "third" neutralizing antibody (in some list) that binds to the hemagglutinin protein, and by the "second" neutralizing antibody that binds to the neuraminidase protein.

Influenza virus only causes acute infections. It is unable to establish a latent or a chronic infection. Consequently, this virus must be passed to new recipients during the week or so while an infected person is still contagious. However, because it is spread efficiently, has a short infectious period, and immunizes infected individuals against a subsequent attack, we might expect that influenza virus would quickly run out of potential human recipients. So how does influenza virus evade the host immune system and thrive in the human population? It turns out that the virus uses variations of the old "bait and switch" routine to ensure that it always has susceptible humans to infect. Here's how it works.

There are a limited number of sites on the hemagglutinin and neuraminidase molecules to which neutralizing antibodies can bind, and since the collection of antibodies each person makes is different, some individuals will produce antibodies that recognize one or more of these sites, but not others. Because the polymerase that copies influenza RNA is error prone, the genetic code of the virus "drifts" as mutations are introduced during copying. Consequently, if, while infecting one person, the virus mutates so that one or more of the antibody binding sites changes, other individuals—who depend on neutralizing antibodies that bind to these mutated sites for protection against the original strain—may no longer be immune to the mutated strain. In essence, the influenza virus offers the immune system one strain to defend against, and then when this "bait" has been taken, it uses its capacity to mutate to "switch" to another strain which the immune system has never seen. The effect of this "antigenic drift" is that individuals with "outdated" antibodies provide a pool of individuals who, although they are immune to the original virus, can be "reinfected" by a mutant strain which arose during infection of another individual.

All three types of influenza virus use antigenic drift to evade the immune system, but influenza A virus has an additional trick up its sleeve. Type A (but not B or C) influenza virus can infect birds and, in particular, waterfowl. In fact, genetic analysis suggests that all human strains of influenza A virus were originally avian (perhaps duck) viruses that, by mutation, acquired the ability to infect humans. One of the intriguing features of this "adaptation" is that whereas type A influenza causes an acute respiratory disease in humans, in ducks, influenza virus causes an infection of the digestive tract that is spread by the fecal-oral route. Prior to beginning their migration at the end of summer, wild ducks gather on lakes in Canada. There, infected ducks defecate, and the virus, which is excreted at high concentrations in the feces, is ingested by young (previously uninfected) ducks who drink the contaminated lake water. Because the number of ducks is great, and infection is so efficient, wild ducks like these represent a large reservoir of novel strains of influenza A virus.

You may be wondering how influenza virus can survive in the acidic conditions of the duck digestive tract and cause an intestinal infection. The answer is that duck versions of influenza A virus are less sensitive to acid conditions than are human strains of flu. Indeed, when human influenza A virus was fed to ducks, no intestinal infection was produced. This illustrates the concept that only slight changes (due to mutations) in the genetic information of a virus can dramatically alter the virus' route of entry and the resulting pathology. This possibility for change has important implications for the emergence of "new" viruses—a subject we will discuss in another lecture.

Humans are rarely infected <u>directly</u> by avian influenza A viruses, because these bird viruses usually reproduce poorly in human cells. However, a pig can be a host to both avian and human influenza A viruses, and occasionally viruses of both origins infect the same pig cell. When this happens, new viruses may be produced that are part human virus and part bird virus. This is possible because the influenza A genome is made up of eight segments of RNA, each of which could be contributed either by the human virus or by the bird virus. Most of these hybrid viruses are not infectious and represent an evolutionary dead end. Sometimes, however, this "mixing in the pig" results in a hybrid virus that can successfully infect humans. If such a hybrid includes the bird gene segment that encodes a novel hemagglutinin protein, there is a chance that none of the neutralizing antibodies produced during earlier flu infections in humans will recognize the bird hemagglutinin. The probability that a re-assortment of viral gene segments will take place according to this scenario is greatest in places where large numbers of pigs, ducks, and humans are in close contact—for example, in Asia.

Whereas antigenic drift results when the error-prone RNA polymerase makes relatively <u>small</u> changes in the genetic code of the virus, the swapping of bird or animal gene segments for human gene segments can result in a <u>dramatic</u> change in the viral genome called "antigenic shift." The availability of a pool of novel avian and pig sequences that can be dipped into to create a new human strain increases the probability that there will always be nonimmune humans available for infection. This also makes it very unlikely that a human vaccine could ever eradicate influenza A virus. In contrast to type A influenza, type B and type C viruses do not have the mechanism of antigenic shift available to them, because neither virus has a nonhuman host which can produce a strain that infects humans.

Influenza-Associated Pathology

Now that we have discussed how influenza viruses reproduce, spread, and evade host defenses, we should be able to predict what the pathological effects of an influenza virus infection will be. Our first clue is that because of its reproductive strategy (involving double-stranded RNA and a viral envelope), influenza virus-infected cells produce large amounts of interferon. And many of the typical flu symptoms—fever, muscle aches, headaches, and fatigue—are caused by interferon. Clearly, causing these symptoms is not something the virus does "on purpose." It simply has no other choice: Production of interferon is an unintended consequence of the way the virus evolved to reproduce.

So the large amount of interferon made during an influenza infection produces the classic flu symptoms that we all know too well. But what about the cases of pneumonia that sometimes result from an influenza infection? Why is pneumonia a consequence of the way influenza virus solves its three problems? Influenza virus infects both resting and proliferating cells in the upper and lower respiratory tract, and as a result of the infection, these cells are killed. One of the consequences of this cell destruction is the sore throat, cough, and hoarseness generally associated with a flu infection. More importantly, however, during a flu infection, large areas of ciliated epithelial cells can be destroyed. Although these dead cells are replaced when undamaged epithelial cells proliferate, in the interval between killing and replacement the "paddles" of these epithelial cells are stilled. As a result, infected individuals become susceptible to superinfection by other pathogens. Indeed, most cases of influenza-associated pneumonia are due to superinfecting bacteria which have easy access to the damaged airways. The killing of cells in the lower respiratory tract (e.g., cells of the bronchioles and alveoli) and the associated inflammation can also interfere with oxygen exchange, narrow the airways, and decrease pulmonary function. As a result, preexisting respiratory diseases such as asthma and cystic fibrosis can be exacerbated. In some cases, the cell killing and inflammation can lead to viral pneumonia.

So all of these symptoms and diseases—from a sore throat to pneumonia—are the consequence of the virus' "decision" to reproduce in a style that induces interferon production and that results in the destruction of cells that line both the upper and lower respiratory tract. However, we shouldn't judge this virus too harshly for making this choice. After all, the virus must reproduce quickly and efficiently to stay ahead of host defenses, and it must elicit a cough to facilitate its spread. Killing the cells it infects after turning them into virus factories is just the virus' way of solving its problems. And a virus has gotta do what a virus has gotta do.

Actually, from the host's perspective, the fact that influenza destroys the cells it infects is not all bad, since extensive cell killing insures that the immune system will be mobilized. Indeed, a strong adaptive immune response (i.e., B cells and T cells) is key to controlling an influenza infection. Because both the innate and adaptive systems respond so vigorously, an influenza infection is simply an annoyance for most otherwise-healthy people. For infants, however, whose immune systems are immature, there is an increased risk for a more serious outcome. At the other end of the age spectrum are elderly people whose immune systems are on the decline, and who frequently have respiratory deficiencies that can be exacerbated by an influenza infection. Over 80% of the deaths associated with influenza occur in people over the age of sixty-five.

All three types of influenza virus use the bait and switch strategy of antigenic drift to evade immune extinction. Antigenic drift results from the error-prone copying of viral RNA, and is responsible for the local outbreaks (epidemics) of influenza infection that occur every year or two.

Type A influenza virus (but not B or C) can also "shift" antigens to evade the host's immune defenses. The antigenic shift evasion strategy stems from the virus' ability to dip into the bird or pig reservoir of RNA gene segments to produce a hybrid virus strain to which no human on earth is immune. Antigenic shift is responsible for global epidemics (pandemics) which occur every decade or two—with serious consequences for humans whose immune systems are unprepared for this bait and switch tactic. The 1918 "Spanish" flu pandemic, which killed over twenty million people worldwide, is an excellent example of the devastation that can result from an antigenic shift.

RHINOVIRUS—A VIRUS THAT SURRENDERS

The average American suffers from a rhinovirus infection about once per year, and roughly half of all cases of the "common cold" are caused by this virus. So rhinovirus would make our top twelve list just on this basis alone. In addition, the contrasts between the lifestyles of rhinovirus and influenza are truly remarkable. Although these two viruses are spread in the same way, they have devised very different strategies to solve their problems, resulting in markedly different pathological outcomes.

Viral Reproduction

Like influenza virus, rhinovirus has a single-stranded RNA genome, but there the similarity ends. Influenza uses receptor-mediated endocytosis to gain entry into the cells it infects. In contrast, the binding of rhinovirus to its cellular receptors results in the destabilization of the virus particle, and the release of its genome directly into the cell's cytoplasm.

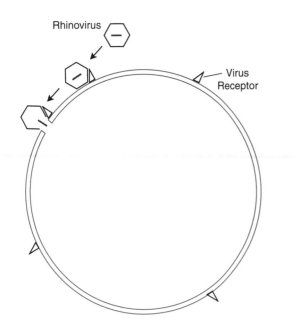

Whereas the influenza virus genome consists of multiple segments of negative-strand RNA, the rhinovirus genome is a single piece of positive-strand RNA. This means that rhinoviral RNA comes into the cell ready to be translated. The product of this translation is a long protein (the "polyprotein") which almost immediately cuts itself into smaller pieces to produce the various viral proteins. One of these proteins, the viral RNA polymerase, makes complementary copies (cRNAs) of the original viral RNA—and then makes many complementary copies of these negative strands to produce the new, positive-strand viral genomes (vRNAs).

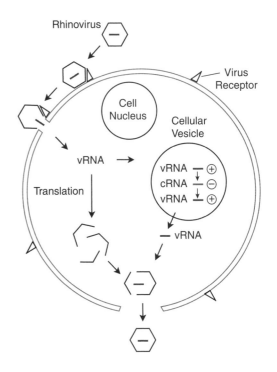

You will notice that in contrast to influenza virus, which replicates within the cell's nucleus, all the action associated with rhinovirus replication takes place within a vesicle located in the cell's cytoplasm. In fact, almost all human viruses with RNA genomes replicate in the cell's cytoplasm, influenza virus being a notable exception.

The new rhinovirus genomes are enclosed in a protein shell (capsid) constructed from virus-encoded proteins, and the newly minted rhinoviruses burst from the infected cell, leaving it dead or dying. This is in contrast to influenza virus, which picks up an envelope as it buds from the infected cell.

As we discussed, influenza virus uses a "cap stealing" gambit as a way of biasing protein synthesis in favor of the virus. Rhinovirus uses a different trick to get the cell's attention. Rhinovirus RNA contains a special initiation sequence that allows ribosomes to begin translation of viral messenger RNA without the benefit of the "cap" structure that normally is needed to tell ribosomes where to begin their work. And to take full advantage of this internal ribosome entry site (IRES), the virus does something really nasty: Rhinovirus encodes a protein that disrupts normal, cap-dependent initiation, effectively shutting down protein synthesis from capped, cellular mRNAs. So although rhino and influenza viruses both recognize the need to focus the biosynthetic machinery of the infected cell on producing viral proteins as opposed to cellular proteins, the strategies they use to achieve this end are very different.

Because rhinovirus is so adept at taking over the host cell's biosynthetic machinery, the virus reproduces very quickly. In fact, it takes only about eight hours for the rhinovirus to reproduce, and for thousands of new rhinoviruses to be made.

Virus Spread

Like influenza, rhinovirus is spread mainly by coughs and sneezes, but there is a subtle difference in the cells these viruses target. Because of the structure of its protein capsid, rhinovirus is most infectious at temperatures that are somewhat below normal body temperature. In fact, rhinovirus' favorite temperature is about 91° F, which just happens to be the temperature found in parts of the nose and upper airways. This is in contrast to influenza virus' preference for 98.6° F—the temperature of the lower airways. So because rhinovirus and influenza have chosen coats which are structurally very different (a capsid vs. an envelope), these viruses tend to infect different regions of the respiratory tract. Nevertheless, like influenza virus, rhinovirus causes only acute infections.

Evading Host Defenses

When respiratory viruses enter their hosts, they encounter a blanket of mucus that coats the respiratory tract and the ciliated epithelial cells whose "paddles" move this mucus. Interestingly, the "mucociliary escalator" actually runs in two directions. There is an "up escalator" that brings mucus up from the lower respiratory tract, and there is a "down escalator" that moves mucus downward from the nasal cavity. In this way, invaders in all parts of the respiratory tract are swept in the direction of the throat to be swallowed or coughed up. Influenza viruses heading toward the lower airways must "swim" against the upward current of mucus. In contrast, rhinovirus actually uses the down escalator to catch a free ride to the interior of the nasal cavity where the temperature is just right for optimal infection. Of course, those rhinoviruses that ride the escalator too far may be swallowed and subsequently destroyed by acids in the stomach. Indeed, rhinovirus' protein coat (capsid) falls apart at low pH—a feature that helps prevent rhinovirus from causing intestinal infections.

Influenza virus is a major interferon inducer which has evolved mechanisms that at least partially block the effects of this interferon on influenza-infected cells. Rhinovirus takes a different approach in dealing with interferon. This virus interferes with the production of interferon by disrupting the cellular system used to transport interferon out of rhinovirus-infected cells. As a result, rhinovirus-infected cells produce much less interferon than do influenza virus-infected cells. And because a rhinovirus infection doesn't generate much interferon, this virus hasn't invested heavily in ways to evade the interferon warning system. In fact, if you're willing to live with the side effects, you can probably avoid ever getting a rhinovirus infection by sniffing interferon alpha every day. This type of prophylaxis won't work for influenza virus, however, because that virus has evolved clever ways of protecting itself from the effects of interferon.

During a flu infection, it is the adaptive immune system (B and T cells) we have to thank for defending us. In contrast, the immune system's defense against a rhinovirus infection is mediated almost entirely by the innate immune system (e.g., complement, professional phagocytes, and natural killer cells). In fact, the innate system does such a great job that most rhinovirus infections are over in just a few days—long before the adaptive immune system can really get cranked up.

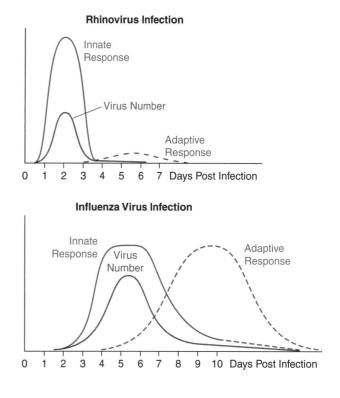

Because rhinovirus "surrenders" so soon after infection, the adaptive immune system usually doesn't become fully activated. As a result, neutralizing antibodies generally are not made in sufficient quantities to protect against a subsequent rhinovirus attack—even by the same strain. So by using this "attack and quickly surrender" tactic, rhinovirus is actually able to infect the same hosts, over and over. In addition, because rhinovirus surrenders just a few days after infection, the symptoms caused by a rhinovirus infection are usually very mild. So this clever virus makes you just sick enough to cough, sneeze, and spread the virus efficiently—but not so sick that you stay home in bed!

Because rhinovirus RNA is all one piece, and because no non-human reservoir of rhinovirus genomes exists, the antigenic shift evasion strategy employed by the segmented influenza A virus is not available to rhinovirus. However, the rhinovirus RNA polymerase is error prone, so like influenza, rhinovirus can use antigenic drift as an additional mechanism to stay one step ahead of the adaptive immune system. In fact, as a result of antigenic drift, over one hundred different strains of rhinovirus are currently circulating in the population. So by using a combination of two evasion strategies—"attack and surrender" plus antigenic drift—rhinovirus deals so effectively with the host defenses that, in terms of the number of people infected each year, it is the world's most "popular" virus.

The Pathology

Influenza virus fights tooth and nail to evade host defenses until the virus is finally overcome by the adaptive immune system. In contrast, rhinovirus reproduces rapidly, kills relatively few cells in the upper airways (focal infections), and quickly surrenders to the innate system. Because of this hit and run lifestyle, the symptoms associated with a rhinovirus infection are mainly due to upper respiratory tract inflammation caused by the innate immune system's strong reaction to the virus attack. This inflammation triggers the sneeze reflex, insuring that the virus will be transmitted efficiently to a new host. However, this sneezing can also spread the infection up into the Eustachian tubes and into the sinuses, causing middle ear infections and sinusitis.

The tissues of the upper respiratory tract are richly provided with capillaries that lie very near the surface. These capillaries function as heat exchangers which, by transferring heat from the blood, warm the room-temperature air as it makes its way toward the lungs. However, this extensive capillary system is also a major target for the cytokines and other inflammatory mediators (e.g., histamine) given off during the innate system's defense against a rhinovirus attack. These mediators cause the capillaries to leak, and the fluids that escape produce the "runny nose" that gives this virus its name (*rhinos* is Greek for nose). Some of this fluid is trapped in the tissues, and causes them to swell, constricting the airways. This congestion can contribute to middle ear and sinus infections by interfering with the normal drainage from these areas. And

the actions of these same inflammatory mediators can exacerbate asthma and chronic bronchitis.

Interestingly, activated macrophages responding to a rhinovirus infection produce interleukin-1, a cytokine that can trigger a low-grade fever. Because rhinovirus is sensitive to elevated temperatures, this increase in body temperature is thought to be one feature of the innate response that is helpful in controlling the spread of the virus.

The typical "flu-like" symptoms of fever, muscle aches, fatigue, and headaches, are much milder in a rhinovirus infection than in an influenza infection. This is because these symptoms are caused in large part by interferon, and rhinovirus-infected cells produce relatively little interferon. In addition, because rhinovirus reproduces best below the core body temperature, rhinovirus infections usually are confined to the upper respiratory tract. As a result, rhinovirus infections rarely result in pneumonia, one of the more common complications of a flu infection.

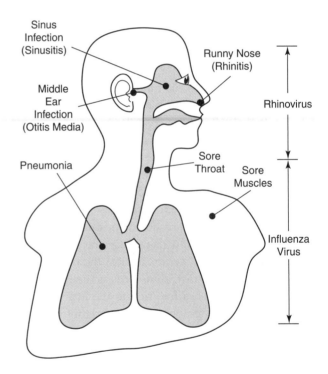

MEASLES—A TROJAN HORSE VIRUS

The final respiratory virus in our Parade is the measles virus. Although measles, flu, and rhinoviruses all enter the body by inhalation, influenza and rhinovirus infections are contained within the airways. In contrast, measles virus can cause a systemic infection of cells throughout the whole body. This illustrates the important point that even viruses which use the same entry portal can cause very different types of infection with very different pathological consequences—all because they have evolved different strategies to solve the problems of reproduction, spread, and evasion.

Viral Reproduction

Like influenza virus, measles virus is a negative strand, RNA virus, but unlike influenza, the measles genome is not segmented—it is a single piece of RNA. Structurally, measles resembles influenza virus in that the measles genome is coated with virus-encoded proteins, and enclosed in a cell-derived envelope. On the other hand, measles uses an entry strategy that has similarities with that used by rhinovirus. After binding to its cellular receptor, the measles virus envelope fuses with the cell membrane, releasing the viral RNA and the proteins that coat it into the cell's cytoplasm where replication takes place.

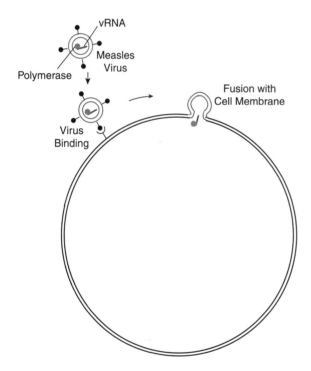

The measles virus RNA polymerase is one of the proteins that remains associated with the viral genome after it enters the cell, and the virus uses this enzyme to make a positive-strand copy of the viral RNA. However, instead of making one long transcript, the viral polymerase stops at specific sites along the RNA, releases the RNA that has been created, and then restarts to produce a total of six, short, positive-strand mRNAs. These are then translated by cellular ribosomes to produce the various measles proteins. The question now, of course, is how does the virus manage to make more full-length, negative-strand RNA molecules to use as genomes for new viruses. Producing complementary copies of the six mRNAs would yield strands of the correct polarity, but these short pieces of RNA would somehow have to be stitched together to make the long, negative-strand genome. You see the problem.

I guess measles virus could have evolved a way of sewing short pieces of RNA together, but instead, this clever virus came up with a much tidier way to produce its genomic RNA. By the time the virus is ready to start packaging new genomes, a large number of viral proteins have been produced. Some of these proteins bind to the original, full-length, negative-strand RNA and mask the signals that formerly instructed the polymerase to stop and restart. As a result of this masking, the polymerase is able to produce a full-length, positive-strand RNA molecule—which can then be copied many times to make new, negative-strand viral genomes. All this action takes place in the cytoplasm of the infected cell, and once the freshly made genomes have been coated with virus-encoded proteins, they bud from the cell surface, picking up parts of the cell membrane as envelopes. These envelopes contain, in addition to the usual cell surface proteins, two viral proteins—the hemagglutinin protein and the fusion protein—which the virus uses to gain entry into its next host cell. Here is an electron micrograph showing measles viruses budding from an infected cell.

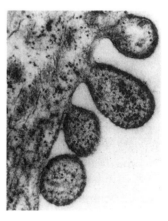

So although measles, flu, and rhinovirus are all RNA viruses that attack the airways, these three viruses have evolved very different ways of reproducing.

How Measles Virus Spreads

The identity of the cellular receptor for measles virus is still a bit controversial. In fact, it appears that this virus may have several different receptors. One of them is a protein called CD46, which is found on the surface of almost every cell in the body. Its normal function is to help defend our cells against attack by the complement system. Using such a ubiquitously expressed protein as its receptor would make good sense, because it would give the virus a wide range of cells to which it could bind.

Like influenza virus and rhinovirus, measles virus is transmitted by the cough or sneeze of an infected individual, and like these other two respiratory viruses, measles virus' sensitivity to acidic conditions precludes its entering the body via the digestive tract. However, in contrast to flu and rhinovirus—which within a few days after infection produce large quantities of virus, cause typical cold or flu-like symptoms, and trigger the cough and sneeze reflexes which efficiently spread these viruses—measles virus usually causes no disease symptoms until about ten days after infection. This period of "silence" results because the innate immune system deals so effectively with the initial measles attack that few airway cells are infected and little new virus is produced. Of course, if this were all there was to the measles story, we wouldn't be discussing this virus at all, for it would have no way to infect new human hosts. As you will see, however, measles virus takes this initial defeat in stride, and uses a "Trojan horse" strategy to spread the virus throughout the body—and eventually back to the airways. Here's how it works.

During the early phases of a measles or influenza virus infection, macrophages, which are battling the infection in the respiratory tract, give off cytokines which dispatch dendritic cells from the battle scene to nearby lymph nodes. Dendritic cells that have been infected by measles virus while on the "front lines" produce viral proteins, and fragments of these proteins are bound to class I MHC molecules, transported to the surface of the dendritic cells, and used to activate T cells in the lymph nodes. It is in this way that the adaptive immune system is alerted when a virus attack has occurred.

Although both influenza and measles viruses infect dendritic cells, the results of these infections are quite different, because an influenza infection of dendritic cells is "abortive." In this incomplete infection, some viral proteins are produced, but little or no infectious influenza virus is made. In contrast, a measles virus infection of activated dendritic cells leads to the production of large quantities of virus. It is a "productive" infection. The reason these two viruses can infect dendritic cells with different outcomes is that measles and influenza viruses employ different reproductive strategies. So whereas dendritic cells have all the "goodies" required for measles virus to reproduce in its particular style, some cellular factors needed for efficient influenza reproduction cannot be supplied by dendritic cells.

The productive infection of dendritic cells by measles virus is critical for the virus' survival, because the viruses produced by the infected dendritic cells infect other cells in the lymph nodes (e.g., activated macrophages), turning these nodes into centers of virus production. Viruses produced in the lymph nodes are then transported via the lymph to the bloodstream—and now the measles virus is back in business. Because the CD46 receptor is so ubiquitous, and because the virus reproduces well in so many different cell types, the viral infection rapidly spreads throughout the body. The disseminated virus infects endothelial cells that line the blood vessels, producing giant, multinucleated cells. These giant cells help spread the infection from blood vessels to nearby epithelial cells, including the epithelial cells which line the airways.

So by using infected dendritic cells as "Trojan horses" to carry the virus into draining lymph nodes, measles virus escapes from the confines of the respiratory tract, enters the lymphatic system, and establishes a systemic infection. In contrast, influenza virus (and probably rhinovirus) cannot productively infect dendritic cells, so it lacks this mechanism for spread beyond the infected airways. As you can imagine, the ability of measles to infect many different areas of the body has a profound influence on the pathological outcome of a measles virus infection. Importantly, because it can establish a systemic infection, measles virus, once banished from the airways, can return there in much greater numbers. It is this second attack which infects and kills large numbers of airway cells, and produces the virus that is spread to susceptible individuals by a sneeze or a cough. Here is a sketch that shows the rough time course of a measles virus infection.

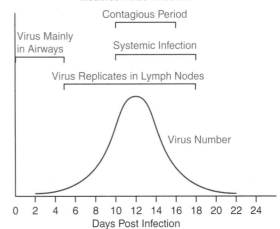

Evasion of Host Defenses

Because measles virus replicates using an error-prone RNA polymerase, we might predict that measles would use rapid mutation as a strategy for evading the attack of virus-specific, neutralizing antibodies. However, this turns out not to be the case. The reason is that, although the error-prone RNA polymerase of the virus does introduce mutations into the viral genome at a high rate, those parts of the viral coat that are targets for neutralizing antibodies cannot mutate without loss of viral function. As a result, those viruses which have mutated so that they cannot be recognized by neutralizing antibodies are no longer infectious.

Because its receptor binding complex is "off limits" to mutations, only one strain of measles virus exists, different strains being defined as viruses that have different binding sites for neutralizing antibodies. Consequently, people who are infected once with measles virus have lifelong immunity to subsequent measles attacks. This dictates that measles must be spread in an unbroken chain in which every new infectee has never been exposed before. In addition, measles virus is so contagious that most humans are infected as young children. As a result, a relatively large population of adults must be living in close proximity in order to turn out enough young children to keep the measles virus alive in the population. In fact, epidemiologists estimate that a minimum population size of about 500,000 is needed. Because localized populations of this magnitude didn't exist earlier than about 6,000 years ago, measles must be a relatively new human virus. Moreover, because other primates (which do have the CD46 receptor and therefore could be infected by measles virus) don't live at such high population densities, humans are the only natural host for measles virus.

So measles can't mutate to avoid neutralizing antibodies in the way that rhino and influenza viruses can. And because measles virus doesn't have a segmented RNA genome, it can't dip into an animal or bird reservoir (as influenza A can) to come up with novel sequences that the immune system has never seen. However, measles does have at least one trick up its sleeve to avoid the antibody defense. On the surface of the measles virus envelope is a "fusion" protein which helps the virus enter its target cells by "gluing" the viral envelope to the outer membrane of the target cell. This fusion protein can also cause measles-infected cells to fuse with underlined, neighboring cells. When this happens, giant cells with multiple nuclei are formed, allowing the measles virus to spread "internally" from cell to cell without ever being exposed to the antibody defense. As a result, resolution of a measles infection is critically dependent on killer T cells which can destroy these infected giant cells.

Although the measles virus replication scheme generates double-stranded RNA molecules, measles infections do not elicit the production of large quantities of interferon. Recent experiments suggest that, like rhinovirus, measles suppresses the production of interferon in the cells it infects. However, the mechanism by which this is accomplished is as yet unknown.

Measles virus also evades host defenses by suppressing the immune systems of infected individuals. This helps buy time for the virus to return to the airway from which it can spread to new hosts. Immunologists speculate that measles-induced immunosuppression stems from the infection and subsequent destruction of dendritic cells and macrophages. Not only does this killing deplete the immune system of two of its most important defenders, but macrophages and dendritic cells also are involved in regulating and directing the actions of other immune system cells. So the killing of macrophages and dendritic cells not only weakens the immune system, it also may lead to a "misdirected" immune response—one in which the wrong class of antibodies is made in response to an infection, or in which too few killer T cells are activated. In addition, there is some evidence that in lymph nodes, infected dendritic cells may actually trigger T cells to commit suicide.

The Pathological Consequences of a Measles Infection

Measles virus is so infectious that without vaccination, essentially everyone will be infected, sometimes with serious consequences. In the United States, vaccination has reduced the number of measles infections to about 300 per year—down from roughly 500,000 per year before vaccines became available. However, because measles vaccinations are not readily available in underdeveloped countries, measles infections still kill over a million people each year worldwide.

Measles virus establishes a systemic infection that eventually brings the virus back in much greater numbers to its starting place in the respiratory tract. There the killing of virus-infected cells triggers a robust inflammatory immune response, and disrupts the function of the mucociliary escalator. As a result, this "second hit" on the airways causes early symptoms which resemble those associated with a rhinovirus infection: fever and a runny nose. Muscle aches and other typical flu-like symptoms are usually absent, because measles virus does not induce the production of large quantities of interferon.

In contrast to both influenza virus and rhinovirus, measles virus must initiate a systemic infection in order to successfully infect the respiratory tract and insure its spread to other humans. And this systemic infection can have pathological consequences that usually do not arise during rhinovirus or influenza virus infections. For example, as a result of the systemic infection, epithelial cells of the eyes are almost always infected. The immune response to this infection can cause inflammation of the mucous membranes that line the inside of the eyelids and the apposing regions of the eyeball ("pink eye" or conjunctivitis) as well as inflammation of the surface of the cornea (keratitis). Interestingly, patients with influenza infections sometimes experience mild conjunctivitis, but this is caused by virus that is carried from the respiratory tract (e.g., by a contaminated finger or a sneeze that blows the virus back up through the tear ducts) rather than by virus that spreads through the bloodstream, as is the case in systemic measles infections.

Cytokines, produced by T cells responding to a measles attack on skin cells, cause the rash that is one hallmark of a measles infection. And infection of the epithelial cells that line the mouth produces the characteristic "Koplik's spots" on the inside of the cheeks. In about half of all measles cases, infection of digestive tract epithelial cells results in diarrhea, nausea, and vomiting. These gastrointestinal problems can be especially serious in developing countries where they can exacerbate malnutrition. One of the most common outcomes of a measles virus infection is immunosuppression—a condition that can persist for more than a year. This immunosuppression can leave the patient open to secondary infections, and is the main reason why over one million children die each year in the Third World as a result of measles infections.

In developed countries, the most frequent life-threatening complication of a measles infection is pneumonia. In children, this is usually caused by superinfecting bacteria which succeed in colonizing the damaged (and relatively defenseless) airways. In adults, measles-associated pneumonia is more frequently caused by the measles virus itself.

In about 0.1% of measles cases, the virus spreads to the brain, and about 15% of the time this results in a fatal demyelinating disease (acute postinfectious measles encephalomyelitis). This condition results when the inflammatory response to the virus infection destroys the myelin "insulation" that normally insures rapid transmission of electrical impulses in the brain. In about one in every 300,000 measles cases, the virus persists in the brains of infected individuals for years, eventually causing a deadly brain disorder, subacute sclerosing panencephalitis.

These serious complications of a measles infection are clearly the unintended consequences of the virus' "decision" to initiate a systemic infection. Indeed, by the time these complications occur, the virus is long gone, having already spread to a new host.

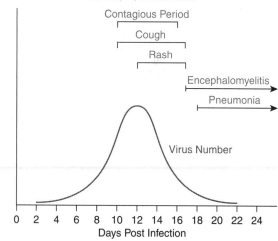

Although the systemic measles infection can have serious consequences, in most cases the adaptive immune system is able to quell the infection and rid the body of the virus. The importance of the immune system in defending against a measles infection is underscored by the fact that patients undergoing immunosuppressive chemotherapy or patients whose immune systems are ravaged by the AIDS virus have a mortality rate from a measles infection which exceeds 40%.

Table 3.1 summarizes some of the important features of the three viruses we have discussed in this lecture.

Table 3.1   Viruses We Inhale

	Influenza	Rhinovirus	Measles
R **E** **P** **R** **O** **D** **U** **C** **E**	Negative, single-stranded, segmented RNA genome	Positive, single-stranded RNA genome	Negative, single-stranded RNA genome
	Enveloped virus	Has single capsid	Enveloped virus
	Error-prone RNA polymerase	Cap-independent translation of viral mRNA	
	Steals caps	Shuts down cellular cap-dependent translation	
	Cytolytic	Cytolytic	Cytolytic
S **P** **R** **E** **A** **D**	Infects epithelial cells in upper and lower respiratory tract	Favors infection of cells in upper respiratory tract	Initially infects cells of respiratory tract; then uses Trojan horse tactic to produce a systemic infection
	Respiratory infection triggers cough reflex	Respiratory infection triggers sneeze and cough reflexes	Spreads when systemic infection returns virus to respiratory tract, and triggers sneeze and cough reflexes
	Causes acute infection	Causes acute infection	Causes acute infection
E **V** **A** **D** **E**	Induces interferon production, but disrupts interferon-induced shutdown of protein synthesis	Blocks interferon production	Blocks interferon production
	Bait and switch: Mutations introduced during viral replication cause antigenic drift	Spreads quickly and then surrenders	Avoids antibody defense by fusing infected and uninfected cells
	Bait and switch: Exchange of gene segments with birds or pigs causes antigenic shift	Mutations introduced during viral replication produce many different virus strains	Immunosuppresses host

Lecture
4

Viruses We Eat

R E V I E W

In the last lecture, we discussed three viruses that enter the body via the respiratory tract: influenza, rhino, and measles. Each of these viruses has an RNA genome, and each uses its own RNA polymerase to make both viral messenger RNA and new viral genomes. This "all RNA" strategy of replication makes these viruses independent of the cellular DNA replication machinery, so epithelial cells which line the airways are fair game for infection, even if they are not proliferating. All three viruses are cytolytic, so they leave their host cells dead or dying once they have used them as factories to produce new viruses. The death of infected cells in the respiratory tract and the accompanying inflammation triggers coughs and sneezes, and microdroplets containing virus particles are expelled into the surroundings where they can be inhaled by other individuals to spread the infection.

The rhinovirus genome is protected by a protein capsid which is assembled within the infected cell. In contrast, influenza and measles virus genomes associate with virus-encoded proteins and then pick up an "envelope" made from patches of cell membrane as they exit. Because of the different ways these viruses "dress," measles and influenza viruses can infect cells deep in the airways where the temperature is 98.6° F. In contrast, the rhinovirus capsid is relatively unstable at this core body temperature, so rhinovirus preferentially infects the upper regions of the respiratory tract where it is cooler. One consequence of this subtle difference in temperature sensitivity is that a rhinovirus infection almost never causes pneumonia.

All three viruses replicate through double-stranded RNA intermediates, and influenza and measles viruses have envelopes. Because of these properties, we would expect all three viruses to induce the expression of large quantities of interferon. This is certainly true of influenza virus, and explains why this virus causes flu-like symptoms. To protect itself from the interferon it produces, influenza virus has evolved mechanisms that reduce the effects of interferon on the cells it infects.

Rhino and measles viruses take a different approach to dealing with the interferon system. Instead of trying to lessen the effect of interferon, both viruses nip the interferon defense in the bud by disrupting interferon production. As a result of this evasion strategy, rhino and measles virus infections are usually associated with fever and a runny nose, caused by the innate system's inflammatory reaction to the virus, rather than flu-like symptoms caused by interferon.

Because all three viruses kill the cells they infect, the innate system, which is on the lookout for "excess" cell death, is rapidly activated. In the case of rhinovirus, the innate response is so potent that after a few days, the virus infection is under control. However, during these few days, rhinovirus reproduces quickly and is spread efficiently to new hosts by coughing and sneezing. In fact, rhinovirus' tactic of "reproduce and surrender" is executed so rapidly that, in most cases, the adaptive immune system is not fully activated, and protective antibodies are not produced.

Measles virus also gets hammered by the innate system in the respiratory tract. To avoid destruction there, this virus infects dendritic cells, and uses a "Trojan horse" strategy to escape the airways and travel to nearby lymph nodes. From these nodes the virus launches a systemic infection in which endothelial and epithelial cells throughout the body are infected. It is this global infection which results in the typical measles symptoms. Importantly for the virus, this systemic infection brings the virus back to the airways in large numbers, making it possible for measles to infect many epithelial cells and to be spread by coughs and sneezes.

Influenza virus neither surrenders (like rhinovirus) nor escapes (like measles virus). Rather, it "stands and fights" in the respiratory system until, like measles virus, it is eventually subdued by a potent adaptive immune response. However, to evade immunological memory and to expand the pool of infectable humans, influenza virus uses two "bait and switch" strategies. During replication, the error-prone viral polymerase introduces mutations into the influenza genome. The result of this "antigenic drift" is that almost every flu virus is genetically different from every other one. Some of these mutations change the viral hemagglutinin protein, so that neutralizing antibodies, which could bind to the original virus and prevent it from reproducing, now become totally useless in preventing reinfection by the mutant virus. When one of these "escape" mutants enters the population, the result can be an influenza epidemic.

To further expand its list of potential infectees, influenza A virus (but not influenza B or C virus) adds an additional twist to the bait and switch routine. The influenza virus genome is made up of multiple RNA segments, and because influenza A virus can reproduce in both birds and pigs, RNA segments from birds or pigs can be picked up by human type A influenza virus. Sometimes these nonhuman sequences encode hemagglutinin molecules that humans have never seen before, and against which they have no protective antibodies. Consequently, the "antigenic shift," produced when bird or pig RNA segments are acquired, can lead to devastating, worldwide influenza pandemics.

In addition to the trick of surrendering before immunological memory has fully matured, rhinovirus also uses its error-prone polymerase to create antigenic drift and evade neutralizing antibodies. In contrast, the part of the measles virus' envelope that is targeted by neutralizing antibodies has a complicated, interlocking structure that cannot "drift" to evade these antibodies without loss of function. As a result, there is only one strain of measles virus, and this strain must be passed in an unbroken chain in which each new infectee has never been infected before.

VIRUSES WE EAT

If the respiratory route of infection represents the easy way in, the digestive tract is most certainly the hard way. The goal of viruses that use this route (the enteric viruses) is to infect epithelial cells that line the walls of the small intestine. To reach these cells, viruses must be able to resist the antiviral defenses present in the saliva, survive exposure to acid pH and digestive enzymes in the stomach, and escape destruction by enzymes that bathe the cells in the intestines these viruses seek to infect. Only a few viruses can do all this. These are their stories.

ROTAVIRUS—AN UNDERCOVER VIRUS

The people who named this virus really got it right. *Rota* is the Latin word for wheel, and that's just what a rotavirus looks like: a wheel with spokes.

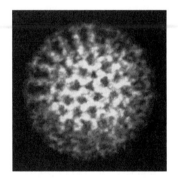

Rotavirus belongs to the reovirus family, and sometimes you hear rotavirus called by its family name. If you classify rotaviruses by the different antibodies that bind to them, there are actually seven different "groups" of rotaviruses. However, group A rotaviruses are the ones that cause most rotavirus-associated diseases in humans, so we'll limit our discussion to this one group.

Viral Reproduction

Rotavirus is rather unusual in that its genome is made up of eleven segments of double-stranded RNA, protected by not one, not two, but three concentric protein shells (capsids). All viruses face the problem of shedding their protective coats during infection, but for a virus with three coats, you'd expect this to be a particularly difficult maneuver. Rotavirus, however, has figured out some very clever ways to deal with this issue.

The outer capsid of the rotavirus includes two different proteins, VP4 and VP7. The VP7 proteins are the primary building blocks of the outer shell, with VP4 proteins sticking out from this shell like spikes.

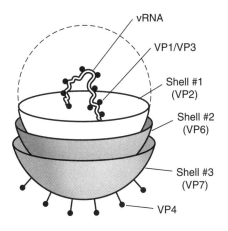

Although the details of the mechanism rotavirus uses to enter its target cells are still a little sketchy, it is clear that both VP4 and VP7 play important roles in this process. The current thinking is that a rotavirus is taken into a cell and enclosed in a compartment called an endosome, which is bounded by cellular membranes. It's worth noting, however, that when biologists don't understand what's going on, they usually give the place it happens the suffix "-some"—and this is no exception. In any case, it is believed that within the endosome, the calcium ion concentration starts out being the same as in the environment outside the cell. Then, VP4 or VP7 or both somehow punch holes in the endosome, allowing the calcium ion concentration surrounding the virus to fall (about ten-fold) to the calcium ion concentration of the cell's cytoplasm. It is this reduction in calcium ion concentration which allows the virus to shed its outer coat. Recent evidence indicates that in the intact virus coat, VP7 is found in "trimers" that consist of three identical

VP7 molecules "bundled" together. When the calcium concentration drops, these trimers fall apart, "unbuttoning" the overcoat. One interesting aspect of this uncoating is that neutralizing antibodies bind to the trimeric form of VP7. Virologists speculate that this binding may prevent the virus from shedding its coat by "clamping" the VP7 trimers together. If true, this would be an excellent example of a mechanism by which an antibody can neutralize a virus infection without acting to prevent the virus from binding to its target cell.

Before VP4 can work its magic, it must be cut by a protease to produce smaller, active forms of the protein. If this cleavage doesn't take place, the rotavirus gets stuck in its coat and is destroyed by the cell it is trying to infect. What's interesting is that rotavirus' target cells, the villus epithelial cells that line the intestine, are bathed in proteases. The "day job" of these proteases is to help cut proteins in the food we eat down to size so that they can be taken up by the body. However, one of these enzymes, trypsin, is the very enzyme rotavirus needs to cut VP4. So a rotavirus infection actually <u>requires</u> the function of a digestive enzyme that would destroy most other viruses daring to enter the small intestine! In effect, rotavirus uses what is normally a barrier defense—the proteases present in the intestine—to prepare the virus for entry.

At this point, the virus has removed its "overcoat," so there are only two coats to go. Now, it might seem that the virus could just go ahead and shed its other two coats, and use its RNA polymerase to copy each strand of its double-stranded RNA. In this way, it could produce both the single-stranded viral mRNA it needs to encode its proteins and the double-stranded RNA required for new genomes. But no. The virus is too smart to do it this way.

In the aqueous environment of the cytoplasm, the two strands of viral RNA would be so tightly bound together that prying them apart to allow the polymerase to work would be very difficult. Double strands of DNA are rather easy to part, DNA-RNA hybrid double strands are more difficult to separate, and the two strands of a relatively long, double-stranded RNA adhere like a tick to a dog. To circumvent this difficulty, rotavirus does something rather ingenious: It uses its RNA polymerase (which is packaged inside the virus particle) to transcribe viral mRNA <u>while the viral genome is still within the protective environment of its double capsid.</u> The single strands of viral mRNA produced by the polymerase are then spit out into the cytoplasm through holes in the double capsid—something like this:

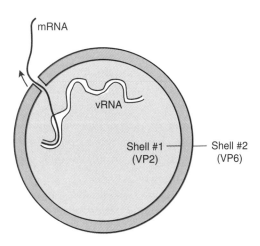

In this illustration I've shown only one viral RNA segment. In reality, all eleven segments are transcribed in this way, and many pieces of single-stranded mRNA come sprouting out of the double capsid at the same time. This strategy works well because inside its double capsid, the environment is not aqueous, so the two complementary RNA strands can be separated more easily to allow the polymerase to do its job.

When it comes time to package new viral genomes, the eleven segments of viral mRNA needed for a complete genome are rounded up and cloaked with proteins (including viral polymerase molecules) that form the inner coat of the new virus. Then the viral polymerases make complementary copies of each of the gene segments to yield the double-stranded RNA viral genome. After genome synthesis is complete, two more protein coats are added, and newly made rotaviruses exit the infected cell, leaving it dead or dying. Because viral mRNA is produced while the virus still has two coats on, and because the double-stranded genomes of new viruses are constructed only after the virus has put on its first coat, rotavirus can be said to "replicate under cover."

Viral Spread

Because of its three protective coats, rotavirus is able to thrive in the harsh environment of the digestive tract. And the requirement for intestinal enzymes to facilitate the "unbuttoning of the virus overcoat" helps explain why rotavirus doesn't establish infections in other parts of the body—areas where these enzymes are lacking. The favorite targets for a rotavirus infection are the columnar epithelial cells at the tips of the villi that line the intestine. Because this is a cytolytic virus, a rotavirus infection leaves the infected villi looking like they have been chewed off. Why the rotavirus prefers the cells at the very tips of the

villi isn't clear. Perhaps these are the easiest cells for the virus to grab onto as it passes by, or maybe it has to do with the fact that the cells at the tips are the most "mature" epithelial cells in the villi. This maturity (whatever that means) may provide an environment within the cell that is especially well suited for rotavirus reproduction.

Although relatively few intestinal cells are attacked during a typical rotavirus infection, these infected cells crank out so much virus that the stool of an infected person can contain as many as one billion viruses per milliliter. In addition, rotavirus remains infectious suspended in water, so the virus can be spread in a contaminated water supply. Because as few as ten rotavirus particles can initiate an infection, it's easy to understand how this virus spreads so efficiently by the fecal-oral route.

Rotavirus is especially contagious among young children. This makes sense: Young children produce feces almost continuously, and they like to put their mouths on everything in sight. Indeed, it is the rare four-year-old who has not been "visited" by this virus. Up until about three months of age, babies who have been breast fed are at least partially protected against a rotavirus infection by maternal antibodies. In this age group, infection is frequently asymptomatic, but some virus can still be produced and shed in the feces. Although maternal antibodies can protect against disease, these antibodies also can prevent babies from becoming immunized by the virus infection. This is because the virus is usually subdued by the inherited immunity before the child's own immune system can mount a vigorous enough immune response to generate memory cells. Once this "passive" immunity from mom "wears off," children become susceptible to infection. Consequently between the ages of six and forty-eight months, most children are infected with rotavirus.

Although rotavirus infections are observed year round, an interesting feature of rotavirus infections is that there is an annual rotavirus epidemic which spreads like a broad wave across North America—starting in Mexico and ending up in the northeastern United States. All sorts of theories have been advanced to explain this wave-like spread, but no single explanation has proved convincing. This raises an interesting point: If rotavirus infections occur mainly once a year, where is the virus hanging out during the "off season"? Rotaviruses can infect many different animals and birds, but no animals have been identified that can efficiently pass type A rotavirus to humans. So if there is an animal reservoir, it has yet to be discovered. It is more likely that rotavirus just "hibernates" in dried stool until conditions are right for it to spread again. Indeed, because of its triple coat, rotavirus is resistant to dehydration, a condition that is lethal to less-well protected viruses. So rotavirus is perfectly adapted for lying dormant until the next kid with an appetite for feces happens by.

Evading Host Defenses

Double-stranded RNA is a potent trigger for interferon production, so we might expect that infection with a double-stranded RNA virus like rotavirus would lead to the production of huge amounts of interferon. However, when rotavirus enters a cell, its double-stranded RNA genome is protected from "view" by its two inner coats. Transcription of viral mRNA takes place within these coats, and <u>single-stranded</u> viral RNA is squirted out into the cytoplasm. Then, when it is time to produce new double-stranded genomes, the single-stranded RNA is first enclosed in a coat of protein, and only then is the second strand of RNA produced. As a result, double-stranded viral RNA is not readily "visible" within the cell, and consequently, relatively little interferon is produced during a rotavirus infection. So rotavirus' "replication under cover" strategy helps it evade the host's interferon defense, buying time for the virus to reproduce and spread.

Because the infected cells die as a result of a rotavirus infection, the innate system is quickly alerted to deal with the attack, and the adaptive system cranks up to produce protective antibodies. Although these antibodies may play an important role in "mopping up" any residual rotaviruses, by the time the immune system really gets going, most of the rotaviruses either have been killed by the innate system or have exited with the feces. Thus, rotavirus is a "hit and run" virus that replicates quickly, and produces a large number of new viruses which then are quickly swept away in the feces to infect the next victim.

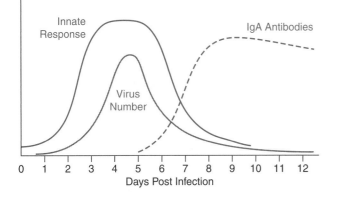

Partly because the immune system only gets a "glimpse" of the rotavirus, and partly because the immune response generated by most mucosal invaders seems to be rather short lived, a rotavirus infection usually does not generate complete immunity to a subsequent infection. What this means is that older children and adults frequently can be re-infected with the same rotavirus strain they fought off as a kid. In many of these reinfections, there is enough immunological memory left to prevent most of the symptoms usually associated with a first infection. Even so, infectious virus can be shed in the feces of asymptomatic individuals, enhancing viral spread.

Like other RNA viruses, the rotavirus genome has a high rate of mutation, and some of these mutations can change the parts of the viral coat that are recognized by neutralizing antibodies. As a result of this antigenic drift, there are always several different rotavirus serotypes circulating in the population.

Viral Pathogenesis

In infants and young children, rotavirus is the major cause of severe inflammation of the intestine (gastroenteritis). Indeed, about one third of the cases of diarrhea severe enough to result in the hospitalization of young children are caused by rotavirus infections. Worldwide, this virus causes nearly a million deaths each year, mostly in underdeveloped countries.

Rotavirus infections frequently result in fever, vomiting, and diarrhea. Early on, it was believed that rotavirus-associated diarrhea was the direct result of the killing of villus intestinal cells by the virus. This killing was thought to cause an imbalance between the fluid-secreting cells in the crypts at the base of the intestinal villi and the mature absorptive cells at the tips of the villi. However, it was later learned that rotavirus usually infects a relatively small fraction of the villus cells in the digestive tract, so cell killing could not be the whole story. Next, it was hypothesized that a toxin encoded by the virus acts <u>directly</u> on intestinal cells (e.g. on crypt cells) to cause diarrhea—but experimental evidence for such a direct-acting viral toxin is lacking. Indeed, the latest experiments suggest that much of the fever, vomiting, and diarrhea that result from a rotavirus infection are caused <u>indirectly</u> through the action of the nervous system. Here's how this is thought to work.

It has long been known that vomiting involves a "reflex loop" which begins when nerves that have their inputs in the walls of the gastrointestinal tract sense that the gut is inflamed. These inflammatory signals are then transmitted by the nervous system to the "vomiting center" in the medulla. If these transmissions are of sufficient strength, neuronal signals are then sent out from the medulla, initiating a rather complex series of events that leads to vomiting. This includes a deep breath, followed by contraction of the diaphragm and the abdominal muscles, and opening of the esophageal sphincter to allow the vomit to escape. This vomit reflex, of course, is a host defense that is designed to help clear the upper gastrointestinal tract of invaders.

Recent experiments in mice now suggest that the diarrhea associated with a rotavirus infection also involves a neuronal reflex loop. In the intestines, there are many activities which must be carefully monitored and coordinated. For example, just the right quantities of digestive enzymes must be released into the small intestine from the pancreas, and just the right amounts of mucus and fluid must be provided by the cells that line the intestines. In addition, the peristaltic contractions of the muscles that line the intestines must be coordinated so that the contents of the intestines are propelled in the correct direction. Reverse peristalsis would not be good. All of these functions are controlled by what neuroscientists call the enteric nervous system. This system has sensors that provide up-to-date information on the environment in the intestines, a processing unit that can make decisions based on this information, and neural outputs that can implement these decisions. Most remarkably, all this can be accomplished without ever sending a signal to or receiving a signal from the brain.

The current thinking about rotavirus-induced diarrhea is that somehow the viral infection stimulates the receptors of the enteric nervous system that are located in the small intestine. If this "I've been attacked" signal becomes strong enough, the diarrhea center in the "gut brain" reacts by generating neural signals that stimulate the crypt cells at the base of the intestinal villi to secrete more fluid into the intestine, causing diarrhea—a defense mechanism meant to help clear invading pathogens from the gut.

So far it is not clear how a rotavirus infection triggers the attack signal. There is some evidence that a rotavirus protein (NSP4), which normally is involved in assembling its coat, binds to receptors on specialized "sensor" cells in the walls of the small intestine. These sensor cells may then react by secreting substances (e.g., serotonin or prostaglandins) that stimulate nearby nerve endings. Another possible (and less elaborate) explanation is that the inflammation caused by the killing

of rotavirus-infected epithelial cells provides the trigger for the vomit and diarrhea reflexes.

Fever, the third common symptom of a rotavirus infection, was originally thought to be due to inflammatory cytokines such as interleukin-1 that travel from the site of infection to the brain, where they can trigger an increase in body temperature. Although this may be true in some cases, recent experiments suggest that inflammatory cytokines can also stimulate nerves in the intestine which then carry the "under virus attack" signal directly to the fever center of the brain. Indeed, the emerging picture is that most of the symptoms of a rotavirus infection—fever, vomiting, and diarrhea—likely result because viral proteins or virus-associated killing of intestinal cells triggers neuronal reflex loops.

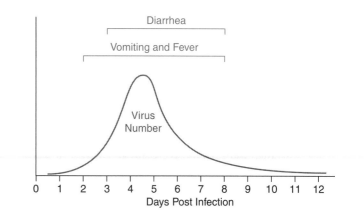

It is important to remember that these symptoms are the <u>unintended</u> consequences of the lifestyle choices made by the rotavirus. Fever, vomiting, and diarrhea are host defense mechanisms, and activating these defenses is certainly of no value to the virus. Indeed, even during asymptomatic rotavirus infections, large amounts of virus can be produced and shed in the feces.

ADENOVIRUS—A VIRUS WITH A TIME SCHEDULE

The second enteric virus we will discuss is the human adenovirus. Actually, human adenoviruses comprise a large family of viruses with about fifty different serotypes. In addition, there are adenoviruses that infect many kinds of birds and animals. Even frogs get adenovirus infections. So far, however, there have been no reports of adenovirus transmission between humans and other species.

Although most human adenovirus serotypes cause infections of the respiratory tract, adenovirus serotypes 40 and 41 specialize in infecting cells that line the intestines. These "enteric adenoviruses" will be our focus here.

Viral Reproduction

The adenovirus genome is a linear, double-stranded, DNA molecule with enough genetic information to encode over thirty proteins—so we can expect this virus to have a number of "bells and whistles" that smaller viruses don't have. In fact, these "extra genes" are what make adenovirus so interesting. In contrast to rotavirus, which has three protein coats, adenovirus only has one. Sticking out of this relatively smooth capsid are "fiber" proteins that give the adenovirus the appearance of a communications satellite.

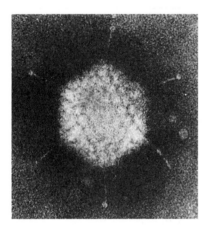

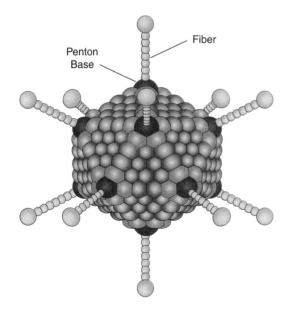

As you might predict, the knobs at the end of the fiber stalks plug into cellular receptor molecules and allow the virus to attach to its target cell. This attachment, however, is not enough to permit the virus particle to enter the cell. In fact, the proteins that make up the base of the fiber stalk, the "penton base," must also plug into co-receptor molecules on the cell surface. Only when this second contact is made can the virus be taken into the cell, enveloped by an endosome.

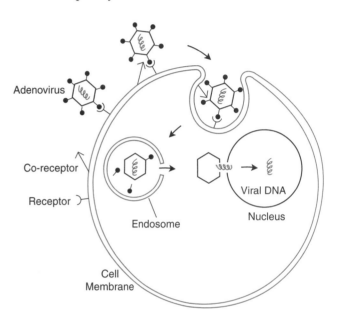

After the virus is partially dismantled by the acidic conditions within the endosome, it escapes into the cytoplasm, and injects its cargo of DNA into the cell's nucleus. You may be wondering: If the virus coat is "partially dismantled" in the acidified endosome, why isn't the virus stripped totally naked by the acidic conditions in the stomach? I could wave my hands and make up a story about how the virus may be protected within a clump of food as it transits the stomach, but the truth is, nobody has a clue as to how adenovirus accomplishes this amazing feat.

Anyway, once the viral DNA reaches the nucleus, cellular enzymes begin to transcribe some of the viral genes into mRNA. Transcription of adenoviral genes proceeds according to a carefully orchestrated program, with certain viral genes being transcribed early after infection, and others being transcribed at later times. It is this ability of adenovirus to function on a strict time schedule that makes it such an effective parasite.

Because both the adenoviral genome and the cell's DNA are linear and double-stranded, the simplest strategy for adenoviral DNA replication would be to just use the cell's DNA replication machinery. After all, why reinvent the wheel? But no. Adenovirus is not content with this "simple" approach. The fact is, no virus exists which uses exactly the same strategy for DNA replication as human cells do—it just won't work for viruses. One major problem has to do with the timing of DNA replication.

When cells proliferate, cellular DNA replication is carefully controlled so that each origin of replication is used only once during a cell division cycle. This insures that each chromosome is copied only once, and that each of the two daughter cells receives one complete copy of the genetic information, and no more. This scheme works just fine for cells, but if viruses used this strategy, only one new virus could be produced per cell cycle. Since viruses make their living by using the cells they infect to make thousands of copies of their genomes, one copy per cell division just won't cut it.

So DNA viruses like adenovirus must somehow uncouple their replication cycle from that of the cells they infect. Adenovirus accomplishes this by using its own DNA polymerase to replicate in a way that is totally different from that of cellular DNA. When cells replicate their DNA, a cellular DNA polymerase moves along one parental strand constructing a continuous complementary daughter strand. At the same time, a DNA polymerase copies the other parental strand. However, because the DNA polymerase only works in one direction, copying this second parental strand results in small pieces of DNA which must subsequently be joined together. So replication of cellular DNA is continuous on one strand and discontinuous on the other.

In contrast, during adenoviral DNA replication, the virus' polymerase makes a complementary copy of one parental strand, displacing the second parental strand. The result is one double-stranded viral DNA molecule plus the displaced single strand. This displaced parental strand is then copied by the viral DNA polymerase to make a second double-stranded molecule. With this scheme, the replication of both strands of viral DNA is continuous.

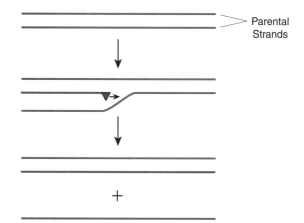

Another difference between cellular and adenoviral DNA replication is that the cellular DNA polymerase requires a short piece of RNA to "prime" DNA synthesis. In contrast, adenoviral DNA replication is primed by a viral protein that binds to one strand at each end of the viral DNA. As a result, replication of adenoviral DNA doesn't have to wait for an RNA primer to be synthesized. And because adenoviral DNA is symmetrical as far as the protein primers are concerned, DNA replication can actually begin at either end. In fact, if there are plenty of replication goodies available, two polymerase molecules can roar along the viral DNA in opposite directions, simultaneously copying both strands.

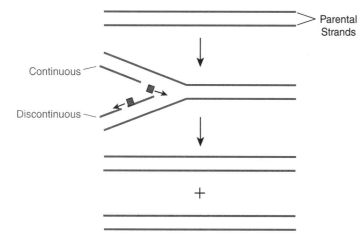

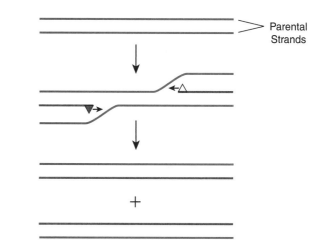

The whole process of adenoviral DNA replication employs not only several viral proteins (e.g., the viral DNA polymerase and primer proteins), but cellular proteins as well. In fact, to replicate its DNA, adenovirus uses many of the supplies that the cell normally would use for replication of its own DNA. These include proteins involved in the copying operation as well as the nucleotide building blocks that are hitched together to form new DNA molecules. This dependence on cellular replication factors raises a potential problem: Many cells the virus would like to infect are not proliferating, and when cells are in a "resting" state, they generally don't keep large quantities of the materials needed for constructing DNA molecules on hand.

To solve the "resting target cell" problem, adenovirus produces, immediately after infection, proteins (the E1A proteins) that kick the infected cell into its DNA replication cycle. When this happens, the cell begins to stockpile the materials the virus will need to make many copies of its DNA. Of course, while these DNA replication supplies are being amassed, it wouldn't do to have the cell replicate its DNA and use them all up. No indeed. So to keep this from happening, the virus shuts down synthesis of cellular DNA. Adenovirus can play this dirty trick because the mechanisms for adenoviral DNA replication and host cell DNA replication are very different.

So adenovirus employs a novel DNA replication strategy which uncouples viral DNA replication from the host cell's replication cycle. This makes it possible for many cycles of viral DNA replication to take place in the time it would normally take the cell to replicate its DNA only once. In addition, the novel replication strategy allows the virus to trick the host cell into making lots of supplies for DNA replication—and then to use them all for viral replication.

After the raw materials required for making DNA have accumulated, the virus begins to make mR-NAs that encode its DNA polymerase and the other proteins required for viral DNA replication. This delay makes sense. There's no reason for the viral DNA polymerase to begin replication until the materials needed for the job are available. Finally, about eight hours after the virus first enters the cell, synthesis of new viral DNA begins within the cell nucleus.

Usurping the cellular DNA replication machinery is not the only "takeover" strategy used by adenovirus. Soon after infection, the virus makes two proteins (the E1B-55K and E4 ORF6 proteins) which interfere with the transport of cellular, but not viral, mR-NAs out of the nucleus. This selective transport of viral RNA insures that viral proteins with enzymatic func-

tions (e.g., the polymerase) are available early on. Then later, when a huge number of proteins must be produced to build thousands of viral capsids, the virus uses another tactic: It rigs things so that translation of viral mRNAs is heavily favored over translation of cellular mRNAs. Focusing the cell's machinery on viral protein production late in infection is important, because construction of each capsid requires over 1,600 protein molecules!

The mechanisms involved in the selective transport and translation of viral mRNAs are not well understood. However, the net result of this takeover is that during the final hours of an adenovirus infection, over 90% of the protein synthesized is viral. So adenovirus not only takes over DNA synthesis in infected cells, it also commandeers the cell's protein synthesis machinery. All this trickery relies on carefully timed expression of viral proteins, and makes adenovirus replication so efficient that a single infected cell can make roughly 100,000 virus particles. This is ten to one hundred times as many virus particles as are produced by cells infected with most other viruses.

Virus Spread

Like rotavirus, enteric adenoviruses spread by the fecal-oral route, and young children are the main targets of enteric adenovirus infections. Interestingly, although many human cells have receptors for adenovirus on their surfaces, enteric adenovirus infections do not spread beyond the digestive tract. This is certainly a good thing. If adenovirus, which produces huge amounts of new viruses during an infection, were to spread throughout the body of a young child, the result would most likely be devastating—both for the child and the virus. After all, a virus that spreads efficiently by the fecal-oral route and kills all the young children it infects would not be a very successful human pathogen.

Although there probably are several factors that contribute to the fortunate containment of enteric adenoviruses within the gastrointestinal tract, one important ingredient is the lack of adenovirus receptors on dendritic cells. If adenovirus could infect dendritic cells, be transported by the infected cells to nearby lymph nodes, and crank out hundreds of thousands of new virus particles, a life-threatening systemic infection would probably result.

Evading Host Defenses

Although each adenovirus-infected cell produces a huge number of new viruses, the process is relatively slow, taking about two days from the time the virus enters until most newly made viruses exit the infected cell. This contrasts dramatically with the rotavirus, which takes only about six hours to turn out a new crop of viruses. This rather leisurely pace of adenovirus reproduction makes adenovirus-infected cells vulnerable to attack by host defenses. Consequently, adenovirus must take effective countermeasures to insure that viral reproduction can be completed before infected cells are destroyed. In fact, roughly one quarter of the adenovirus genome is dedicated to countering host defenses.

All cells have "alarm systems" that alert them when biosynthesis is not going as planned, and which trigger a cascade of events within the cell that leads to the cell's death by apoptosis. This type of quality control is necessary, because a cell's biochemical systems are so complicated that dysregulated cells are a frequent occurrence—and these out-of-control cells can pose a real threat to the human organism (e.g., by leading to cancer). Because adenovirus totally usurps the biosynthetic machinery of the cells it infects, you can be certain that these alarms go off. And if the virus did nothing to prevent it, apoptosis programs would surely be set in motion that would kill the cell and the virus within it long before the virus could reproduce.

At least two adenoviral proteins are devoted to dealing with the host's "suicide" defense against stressed cells. As is appropriate, both are expressed early after infection. One (the E1B-55K protein) blocks transcription of cellular genes that normally would activate the death program. The other (E1B-19K) binds to and inactivates key host proteins involved in initiating apoptosis. These apoptosis inhibitors make it possible for the virus to disrupt normal cell activities without the cell responding to the "something is terribly wrong" alarm.

Suppression of apoptosis during an adenovirus infection is an excellent idea, because a dead cell isn't going to produce much virus. But this suppression also poses a potential problem. Although an adenovirus-infected cell is pretty beat up by the time viral reproduction is complete, unless something is done to really rip the cell open, the newly made viruses will just slowly dribble out. Viruses like rotavirus, which reproduce quickly, actually use apoptosis to help accomplish their "final exit" from the cell. But adenovirus, because it suppresses apoptosis, must make other exit arrangements.

Late in infection, when virus assembly is nearly finished, adenovirus produces a protein aptly named "the adenovirus death protein" (E3-11.6K). This protein acts in a still-mysterious way to burst open the infected cell and allow the 100,000 new viruses trapped inside to come roaring out. The importance of this very late-appearing protein is demonstrated by the discovery that adenovirus mutants which cannot make the death protein take about an extra week to exit an infected cell. So the E1B proteins, which are made early in infection, hold off cell death long enough for viral reproduction to be completed. Then, just when newly made viruses are ready to exit, the ade-novirus death protein delivers the *coup de gras*. Yes, timing is everything.

Another peril that adenovirus must face is the interferon system. It might seem that this virus would not induce interferon production in the cells it infects. After all, adenovirus has no lipid envelope, and its genome does not replicate through a double-stranded RNA intermediate. However, adenoviral mRNAs are transcribed from both strands of the viral DNA genome, and to take full advantage of its coding capacity, the virus allows some of these coding regions to overlap. That way the same stretch of DNA can produce two different mRNAs—one from one strand and a second mRNA from the other, complementary strand. One result of this arrangement of genes, however, is that mRNAs transcribed from these overlap regions have complementary sequences which can base pair to produce long, double-stranded RNA molecules. So by positioning genes opposite each other on its two DNA strands, adenovirus increases its capacity to make proteins. But it does so at a cost—exposure to the interferon defense system.

To protect itself from the interferon it induces, adenovirus produces "decoy" RNA molecules called VA RNA. Normally, in interferon-alerted cells, double-stranded viral RNA binds to the sensor protein, PKR. It is this binding which activates PKR (a protein kinase), and shuts down protein synthesis, ending the viral infection. However, in adenovirus-infected cells, the VA RNA binds to PKR and renders the protein kinase inactive, allowing protein synthesis to continue.

In addition to producing molecules that help the virus cheat death at the hands of the apoptotic and interferon "executioners" within the cell, adenovirus must also defend itself against attacks from outside the infected cell. Indeed, because adenovirus reproduces slowly, adenovirus-infected cells should be attractive targets for destruction by killer T cells. Killer T cells recognize fragments of viral proteins displayed by class I MHC molecules on the surface of an infected cell. Because adenovirus makes so many different proteins in such abundance, there is no way that fragments of some of these proteins won't fit nicely in the grasp of class I MHC molecules and be put on display on the cell surface. Such an advertisement of the infected status of the cell would likely result in its destruction by killer T cells, well before new viruses could be produced.

As a defense against killer T cells, adenovirus has evolved a mechanism that keeps viral proteins from being displayed on the surface of virus-infected cells. Normally, class I MHC molecules are loaded with protein fragments in the endoplasmic reticulum from whence they proceed to the cell surface to display their cargo. However, in adenovirus–infected cells, a viral protein (E3-19K), which is anchored firmly in the endoplasmic reticulum, grabs class I MHC molecules and prevents them from traveling to the cell surface. This "not so fast, Buster" strategy works nicely, because if killer T cells don't see viral proteins displayed on the cell surface, they have no way of knowing that the cell has been infected. Of course, the E3-19K proteins can't snag every single class I MHC molecule as it passes by, so this evasion strategy isn't perfect. But it does greatly decrease the destruction of adenovirus-infected cells (and the viruses inside them) by killer T cells.

But wait! Won't preventing the expression of class I MHC molecules on the cell surface make adenovirus-infected cells excellent targets for attack by natural killer cells? After all, natural killer cells do specialize in destroying cells that don't have class I MHC molecules on their surfaces. Fortunately, the clever adenovirus has evolved ways to deal with at least one of the weapons carried by natural killer cells.

Killer T cells and natural killer cells both have two different ways to kill cells. First, these killers can trigger death by apoptosis when proteins on their surfaces (e.g., FasL) plug into "death receptors" (e.g., Fas) on the surfaces of their targets. To counter this weapon, adenoviruses make a protein called RID that binds to the death receptors on adenovirus-infected cells, removes them from the cell surface, and oversees their destruction. As backup, the adenovirus makes another protein (E3-14.7K) that interrupts the signal from the death receptors—just in case RID misses a few. As a result of the action of these two virus proteins, one of the potent weapons used by both killer T cells and natural killer cells is effectively neutralized.

The second way killer T cells and natural killer cells destroy virus-infected cells involves enzymes called granzymes that these killers "squirt" into their target cells. So far, nobody has discovered a strategy that protects adenovirus-infected cells against this kind of killing. In fact, in all the world of viruses, there is not a single example of a virus that has evolved a defense against granzyme-mediated killing. From the virus' perspective, this is probably a good thing. If the adenovirus were <u>completely</u> protected against killing by killer T cells and natural killer cells, its human hosts would be almost defenseless against the viral infection. Because a dead host is usually not a good host, humans and adenoviruses probably have reached a standoff in which adenovirus has evolved just enough countermeasures to allow it to reproduce efficiently without seriously damaging its hosts.

Human adenoviruses have about fifty different serotypes, defined by neutralizing antibodies which recognize different versions of two of the proteins that make up the viral capsid (fiber and hexon). So it would seem that the adenovirus DNA polymerase must be error-prone. Indeed, although experiments indicate that the viral DNA polymerase has some capacity to "proofread" its work, the adenovirus polymerase certainly isn't nearly as error-free as the cellular DNA polymerase—an enzyme complex that makes only about one mistake per 100 million bases. So adenovirus had another good reason for not using the cellular DNA polymerase to replicate its genome: By using its own polymerase, adenovirus is able to generate antigenic drift, albeit probably not as rapidly as RNA viruses like rotavirus.

Pathogenic Consequences of an Enteric Adenovirus Infection

Enteric adenovirus infections are second only to rotavirus infections as the most frequent cause of infantile diarrhea, and by the age of three, most children have been infected by an enteric adenovirus. Although rotaviruses and adenoviruses are extremely different in terms of their reproductive strategies and evasion tactics, both viruses spread by the same route, infect and kill the same cells, and do not cause disseminated infection. As a result, in the clinic, it is usually impossible to differentiate between an enteric adenovirus infection and a rotavirus infection based on symptoms alone.

Although the symptoms of adenovirus and rotavirus infections are the same—fever, vomiting, and diarrhea—the time courses of these two infections are different. The reason, of course, is that the rotavirus reproduces very quickly, whereas the adenovirus takes its sweet time. Usually about a week elapses between an adenovirus infection and the appearance of any symptoms. This makes sense, because it takes several days for even the first adenovirus-infected cells to begin to produce virus. In contrast, a week post infection, the rotavirus has "left the building," and the symptoms of a rotavirus infection have usually abated. Not only do the symptoms of an enteric adenovirus infection appear later, these symptoms generally last longer than those of a rotavirus infection. This is because, in contrast to the "hit and run" rotavirus, the adenovirus has invested heavily in tactics that allow the virus to evade host defenses for a relatively long time. Eventually, the host's immune system does deal harshly with an enteric adenovirus infection, and the virus is banished (cleared) from the host. And because the adenovirus doesn't "go gentle into that good night," the adaptive immune system becomes fully activated, and immunity to the infecting strain of adenovirus is long-lasting. This is quite different from the situation with a rotavirus infection in which the adaptive immune system only gets a glimpse of the virus, and as a result, immunity to subsequent infections is usually incomplete.

In this lecture we have concentrated on two adenovirus serotypes, 40 and 41, that cause gastrointestinal disease. However, adenovirus got its name because it was first isolated from human adenoid tissue, and many adenovirus serotypes do cause upper respiratory infections. Still others cause childhood pneumonia, and some, as we have discussed, cause gastroenteritis. Interestingly, some adenovirus strains, for example serotypes 4 and 7, can infect both the respiratory tract and the gastrointestinal tract. Immunologists take advantage of the "dual targets" of these two serotypes when they vaccinate army recruits to prevent respiratory infections. That vaccine is made by packaging live adenovirus 4 and 7 in gelatin capsules, which are then swallowed by recruits. Administered in this way, the viruses in the vaccine bypass the respiratory tract, where they would cause acute respiratory disease, and go on to establish an asymptomatic, immunizing infection of the epithelial cells of the small intestine. What's so elegant about this vaccination strategy is that the immunity generated by this asymptomatic intestinal infection protects not only against future intestinal infections, but also against respiratory infections by serotypes 4 and 7.

HEPATITIS A—A VIRUS THAT DETOURS

According to most estimates, over half the population of the United States has been infected with hepatitis A virus, so we certainly need to include this one in our Parade. But what really makes this virus so interesting is that, although it enters its host through the mouth and exits through the anus, just like rotavirus and the enteric adenoviruses, on its trip from top to bottom it takes a "detour" through the liver. It is this detour which makes hepatitis A virus such a successful human pathogen. It is also the main feature of the virus lifestyle that leads to the pathological consequences of a hepatitis A virus infection.

Viral Reproduction

Hepatitis A virus consists of a single piece of positive-strand RNA encased in a single capsid made of protein—just like the rhinovirus. In fact, these two viruses are so similar in their organization that it is widely assumed they reproduce in very similar fashions. However, there is at least one major difference in the ways rhino and hepatitis A viruses reproduce. Whereas rhinovirus shuts down synthesis of host proteins, killing the cells it infects, hepatitis A virus reproduces "gently," making new viruses without causing perceptible damage to its host cells.

Viral Spread

Another important difference between rhinovirus and hepatitis A virus is that the capsid of hepatitis A virus is resistant to the acid conditions in the stomach—conditions that would destroy the rhinovirus capsid. As a result, rhinovirus is a respiratory virus, and hepatitis A virus usually is spread by the fecal-oral route. This illustrates the important concept that subtle changes in viral design can result in major differences in the route of entry of a virus and in the diseases that result from a virus infection.

Hepatitis A is a virus that targets the liver—an organ through which blood flows continuously. So we might predict that this virus would be spread when humans exchange blood or blood products (e.g., during a blood transfusion or when drug abusers share needles). However, hepatitis A virus usually is eliminated quickly by a patient's immune system, and hepatitis A virus never establishes a chronic infection. As a result, the period of time when an infected individual's blood contains hepatitis A virus is generally so short that the probability of spread by blood to blood contact is quite small.

In contrast, hepatitis A virus is well suited to be spread efficiently by the fecal-oral route. After four weeks in dried feces, its infectivity only decreases by about a factor of 100. In addition, the virus can survive for weeks in shellfish, which can concentrate the virus by filtering large volumes of contaminated water. Fortunately for us, hepatitis A virus is sensitive to the concentrations of chlorine commonly used for water treatment (and also to toilet bowl cleaner!).

There is only one known serotype of hepatitis A virus, and infection generally leads to life-long immunity. So hepatitis A virus depends on lax hygiene and a large population of susceptible individuals for its continued existence. Although several types of animals can be infected with hepatitis A experimentally, there is no known natural animal reservoir for this virus.

Evading Host Defenses

As you would expect for a virus which is primarily spread by the fecal-oral route, hepatitis A enters cells in the small intestine. However, infection of these cells has been extremely difficult to document. Moreover, newly made virus is usually not detected in feces until weeks after a hepatitis A infection. This suggests that very few intestinal cells are infected, and that they produce relatively few new virus particles. Now of course, if this were the complete story, hepatitis A virus would be in deep trouble in terms of persisting in the human population. However, this cunning virus has a trick up its sleeve that makes it all work nicely: Hepatitis A virus takes a "detour" to the liver on its way through the digestive tract, and in doing so, evades host defenses long enough to become one of the world's leading pathogenic viruses.

It has been known for years that the primary target of a hepatitis A virus infection is the liver. That, of course, is why it's called a hepatitis virus (*hepato* is Greek for liver). In contrast to its infection of intestinal cells, which is pretty wimpy, a hepatitis A infection of liver cells is robust, producing large quantities of virus. These newly made viruses are released into the bile ducts that drain the liver, and are subsequently emptied into the intestines with the bile. In fact, production of hepatitis A virus by liver cells is so efficient that at the height of a hepatitis A infection there can be 100 million virus particles excreted per milliliter of feces. That's a lot of virus! Although it is clear that most hepatitis A virus infections begin in the intestine, that the virus then detours through the liver, and that newly made viruses are delivered back to the intestine with the bile, the big question has been, "How do the small number of virus particles generated during an intestinal infection ever manage to make their way to the liver and establish an infection there?" Very recent experiments have suggested an elegant solution to this longstanding problem. Here's how it is believed to work.

Keeping watch over our intestines is the mucosal immune system. Although the "rules" that govern this arm of the immune response are not as well understood as those which apply to the immune system in other parts of the body, a picture of how mucosal immunity works is starting to come into focus. It is likely that when hepatitis A virus infects cells that line the intestines, some of the newly made viruses are transported through specialized M cells into the tissues below. These M cells are tasked with sampling the contents of the intestines, and with helping to initiate an immune response to invaders. From the tissues beneath the M cells, the virus is transported to nearby lymph nodes where B cells are activated which can produce antibodies that can bind to the virus. This process takes a week or more while the selected B cells proliferate to build up their numbers, travel back to the tissues underlying the intestines, and begin to pump out antibodies specific for hepatitis A virus. These antibodies are generally of the IgA class, because this class of antibody is especially well suited to defend against intestinal invaders. IgA antibodies can be transported into the intestine itself, bind to newly made viruses, and shepherd them out of the intestine with the feces. And IgA antibodies can also bind to viruses which have invaded the tissues that surround the intestines. However, IgA antibodies have one important weakness—a weakness that hepatitis A virus has learned to use to its advantage.

Other antibody classes (e.g., IgG antibodies) can form a bridge between an invader and a professional phagocyte (e.g., a macrophage), making it easier for the phagocyte to "eat" the invader. However, IgA antibodies perform this maneuver with great difficulty, because phagocytic cells have only low-affinity receptors on their surfaces for IgA antibodies. As a result, IgA antibody-virus complexes are not efficiently "cleared" from the tissues by phagocytic ingestion. Rather, IgA antibody-virus complexes in the tissues are collected by the lymphatic system, poured into the bloodstream, and sent to the liver. Liver cells (hepatocytes) do have receptors for IgA antibodies (asialoglycoprotein receptors), and once connected to these receptors, both the IgA antibodies and their cargo of invaders are taken inside the liver cell for disposal.

It now appears that hepatitis A virus actually uses this IgA "disposal system" to solve its problem of how to infect liver cells. As IgA antibody-virus complexes collected from intestinal tissues travel to the liver for disposal, hepatitis A virus happily goes along for the ride. And when the IgA antibody-virus complexes reach the liver, the virus uses the uptake of these complexes to usher it into the very cells it wants to infect. Once inside hepatocytes, the virus somehow escapes destruction, and begins to reproduce. Newly made viruses are then released by the infected liver cells and sent back into the intestines along with the bile that the liver generates. This return trip to the intestines is a piece of cake for this virus, because the protein coat of hepatitis A is unfazed by the bile salts, which act as detergents and would destroy "ordinary" viruses. But hepatitis A is no ordinary virus!

It gets even better, however. Like politics, immune responses are usually local. They have to be, because the human body is continually under attack on many fronts (e.g., the respiratory tract and the digestive tract) by different invaders (e.g., viruses and bacteria) which require different types of immune responses (e.g., different classes of antibodies). In particular, the mucosal immune system and the IgA antibodies it generates are designed to protect the digestive tract, not the liver. So when hepatitis A virus infects liver cells, the immune response must begin from scratch to activate new B cells which can produce the IgG antibodies and killer T cells that are needed to rid the liver of the virus. And during the week or two while the "liver" immune system is firing up, the virus is reproducing like crazy in infected liver cells. Newly made viruses are also released from infected liver cells into the bloodstream, and these viruses can amplify the infection by entering additional liver cells without the aid of IgA antibodies.

So hepatitis A virus' envasion tactic is to initiate a limited infection in the intestine, travel to the liver, produce copious new viruses in infected liver cells before the liver immune response can crank up, and return these new viruses to the digestive tract via the bile so they can be excreted in the feces. It is this detour through the liver that makes the virus lifestyle work.

Viral Pathogenesis

Because the initial infection of cells in the small intestine is so gentle, the early phase of a hepatitis A infection is uniformly asymptomatic. A week or two into the infection, the virus makes its way to the liver, and viral particles produced by infected liver cells begin to be excreted in the feces. After another week or so, killer T cells, activated in response to the liver infection, start to appear, and the destruction of infected liver cells begins. Because hepatitis A is not a cytolytic virus, this cell killing results from the immune response to the infection, not from the viral infection itself. Fortunately, the number of hepatitis-A-infected liver cells is usually too small to compromise liver function. After all, the liver is a big organ. Consequently, most hepatitis A infections, especially in young children, are asymptomatic.

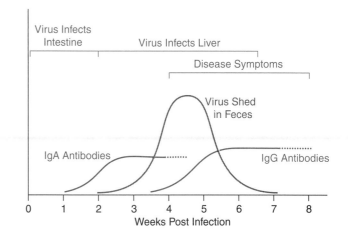

In a minority of hepatitis A infections, mainly in adults, the destruction of liver cells is more extensive, and symptoms characteristic of viral hepatitis begin to appear about four weeks post infection. Certainly the most striking hepatitis symptoms are jaundice and dark urine. When your eyes turn yellow and your urine runs dark, you know something ain't right.

In the human body, about 100 billion aged red blood cells are "retired" each day. These effete cells are rapidly digested by macrophages, and the iron they contain is recycled. However, the part of the hemoglobin molecule that cradles the iron atom cannot be reused, and after it has been processed by the macrophage to form a yellow pigment called bilirubin, it is spit out into the blood or tissues surrounding the macrophage. Because each red blood cell

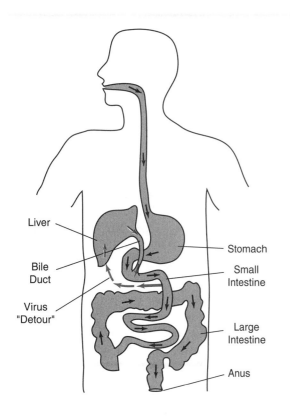

contains so many hemoglobin molecules, and because so many red blood cells are destroyed each day, the huge amount of bilirubin produced creates a major disposal problem. To deal with this, most of the bilirubin is complexed with proteins in the blood (primarily albumin) to make it soluble, and is carried to the liver. There the bilirubin is taken up by hepatocytes, modified, and released into the bile. When the system is working properly, disposal by the liver is so efficient that the concentration of bilirubin in fluids and tissues remains low. However, when large numbers of liver cells are destroyed by the immune response to a hepatitis A infection, this disposal system can be overwhelmed. When that happens, bilirubin concentrations increase dramatically, resulting in jaundice and dark urine.

Although bilirubin is not terribly toxic, jaundice and dark urine are pretty good indicators that the liver is not functioning properly. Because the liver is tasked with detoxifying many other waste products of normal cellular metabolism, jaundice is usually accompanied by symptoms such as malaise, loss of appetite, fever, nausea, and vomiting—symptoms which are caused by inadequate

detoxification of cellular waste products due to liver damage. Fortunately, hepatitis A virus never establishes a chronic infection, the immune system usually requires only a few weeks to eradicate the virus, and liver cells destroyed by the immune response are quickly replaced by the proliferation of healthy liver cells. As a result, symptoms associated with a hepatitis A infection are generally short lived, and this virus rarely causes life-threatening disease. In the United States, for example, there are only about 100 deaths each year associated with hepatitis A infections, and these are mainly in older age groups.

So the detour that hepatitis A virus takes on its way from mouth to anus makes it possible for the virus to cause an <u>acute</u> infection of the liver. This infection is easily dealt with by the immune system before much liver damage has occurred, but after a large number of new viruses have been pumped out with the feces—viruses which can then go on to infect other humans via the fecal-oral route.

Table 4.1 reviews how our three model enteric viruses solve their problems of reproduction, spread, and evasion.

Table 4.1   Viruses We Eat

	Rotavirus	Enteric Adenovirus	Hepatitis A
R E P R O D U C E	Segmented, double-stranded RNA genome	Large, linear, double-stranded DNA genome	Positive, single-stranded RNA genome
	Has three protein coats	Single protein capsid	Single protein capsid
	Replicates "under cover" within its coats	Replicates slowly, using carefully timed plan of viral gene expression	
	Uses digestive enzymes to prepare virus for entry	Takes over control of DNA and protein synthesis in infected cells	
	Cytolytic	Cytolytic	Non-cytolytic
S P R E A D	Infects cells at tips of intestinal villi	Infects intestinal epithelial cells	Initially infects intestinal cells; then infects liver cells
	Fecal-oral spread	Fecal-oral spread	Enters digestive tract; "detours" through the liver; exits with feces
	Causes acute infection	Causes acute infection	Causes acute infection
E V A D E	Hides from interferon system by replicating under cover	Viral proteins hold off apoptosis until replication is complete	Confuses immune system by detouring through liver
	Speedy replication—hit and run; immunity not complete	Virus produces decoy RNA molecules to trick interferon system	
	Antigenic drift produces multiple strains	Interferes with viral antigen presentation by MHC molecules	
		Neutralizes one of natural killer cell weapons	

Lecture 5

Viruses We Get from Mom

R E V I E W

In the last lecture we discussed three examples of enteric viruses—heroic viruses that brave many perils to infect cells of the gastrointestinal tract. First we examined the rotavirus with its segmented, double-stranded RNA genome protected by three protein coats. This virus was contrasted with the human enteric adenoviruses which have non-segmented, double-stranded DNA genomes and only one protein coat. Whereas the rotavirus reproduces very quickly (in only about six hours), adenovirus reproduces slowly, with over thirty different genes being expressed in a carefully controlled sequence that spans a period of several days. Although their reproductive strategies are about as different as two reproduction schemes can be, both viruses are spread by the fecal-oral route, and both viruses infect the epithelial cells of the small intestine, causing fever, vomiting, and diarrhea. Thus, two viruses which have evolved to live wildly different lifestyles can spread by the same route, infect the same cells, and cause the same disease symptoms.

To evade destruction by host defenses, these two viruses use very different strategies. Rotavirus is a "pacifist" which makes use of proteases present in the intestine to help it uncoat, and avoids the interferon defense system by replicating "under cover" within its protective coats. Rotavirus uses the cell's apoptosis defense system to help it escape infected cells, and reproduces so quickly that it usually leaves the infected host before the adaptive immune system becomes fully activated. Because it uses this hit and run strategy, the symptoms associated with a rotavirus infection generally last only a few days, and immunity to a rotavirus infection is usually not complete. Consequently, a person can be infected again later in life by the same rotavirus serotype.

Adenovirus, in contrast, is an "activist." By cleverly timing the expression of its many genes, adenovirus completely takes over its host cell, blocks the cell's apoptotic response to the takeover, thwarts the interferon system by interfering with its activation, and resists destruction by killer T cells and natural killer cells. Adenovirus also produces a "death protein" at exactly the right time to "explode" the infected cell, and facilitate the rapid exit of about 100,000 newly made viruses. Although adenovirus goes down fighting, the adaptive immune system eventually does destroy the virus, and immunity against the infecting adenovirus serotype is long-lasting. Because adenovirus fights to the end, the symptoms associated with an enteric adenovirus infection can persist for a week or more.

Although hepatitis A virus enters by the mouth and exits by the anus—just as rotavirus and the enteric adenoviruses do—the infection of intestinal cells by hepatitis A virus is of relatively little consequence. For this virus to survive, it must infect liver cells. The latest experiments suggest that hepatitis A virus accomplishes this feat in a rather devious manner. When hepatitis A virus infects cells in the small intestine, IgA antibodies are produced which can bind to the virus. These antibodies then travel through the lymph and blood to the liver, carrying their "cargo" of hepatitis A viruses. Liver cells have receptors for IgA antibodies on their surfaces, and normally IgA antibody-virus complexes from the blood are taken into liver cells and destroyed. However, hepatitis A virus has a trick or two up its sleeve, for once inside a liver cell, the virus evades destruction and reproduces efficiently. The newly made hepatitis A viruses then leave the liver with the bile, are poured back into the small intestine, and exit with the feces to infect others via the fecal-oral route. Thus it appears that hepatitis A virus "hitches a ride" to the liver by taking advantage of the immune system's response to its initial infection of cells in the small intestine.

The "detour" that hepatitis A virus takes on its way from mouth to anus makes it possible for this virus to persist in the human population. However, this feature of the virus lifestyle also is responsible for the disease symptoms associated with hepatitis A infections—symptoms that occur when the destruction of infected liver cells by the immune response compromises liver function. Fortunately, hepatitis A virus never establishes a chronic infection, and the virus is usually eliminated by the immune system before much liver damage can occur.

VIRUSES WE GET FROM MOM

In this lecture, we will focus on three viruses which can be spread from mother to child: hepatitis B, hepatitis C, and HTLV-I. Although these viruses have very different lifestyles, they have one important common feature: All are able to establish lifelong, chronic infections. This makes perfect sense for a virus that spreads from mother to child, because the virus must persist in the body of the infected child until she, too, can become a mom.

HEPATITIS B—A DECOY VIRUS

Hepatitis B virus has the smallest genome of any human virus, with only about 3,200 base pairs of genetic information. Compared to adenovirus, which has a DNA genome comprised of roughly 35,000 base pairs, hepatitis B virus is a genetic midget. Yet despite its limited coding capacity, hepatitis B virus is one of the world's most successful human pathogens.

Viral Reproduction

Hepatitis B virus has a circular, DNA genome which is mostly double stranded. This genetic information is protected by a protein capsid plus an envelope. Humans are the only natural hosts for hepatitis B virus, and this virus easily wins the award for "Virus With the Most Bizarre Replication Strategy." To initiate an infection, proteins on the virus envelope bind to unidentified receptors on the surface of the target cell, the envelope fuses with the cell membrane, and the encapsidated genome is released into the cell's cytoplasm. There the viral capsid is removed, and the viral DNA enters the cell's nucleus. The hepatitis B genome has a gap where its DNA circle has only one strand, so when the viral genome reaches the nucleus, it is "repaired" by an unidentified polymerase to produce a completely double-stranded, circular DNA molecule.

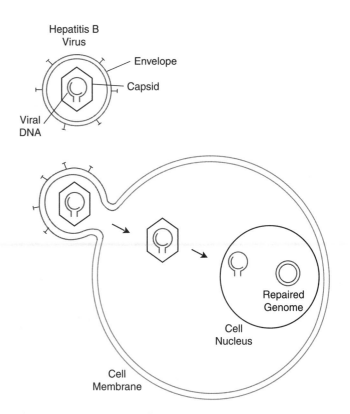

Once the genome has been repaired, the virus uses the cell's RNA polymerase to transcribe one of its DNA strands (the negative strand) into mRNAs of various lengths. These mRNAs are then translated to make the proteins required for viral reproduction. In addition, the cell's RNA polymerase also makes full-length, complementary copies of the negative DNA strand to produce "genomic" RNAs. And here's where things start to get strange. After this genomic RNA is transported to the cytoplasm, the proteins that will form the viral capsid begin to assemble around the RNA. So at this point, it would appear that hepatitis B is about to become a virus with a single-stranded RNA genome. But wait! As the capsid is being constructed, a virus-encoded enzyme (reverse transcriptase) begins to make a complementary DNA copy of the genomic RNA, degrading this RNA once it has been reverse transcribed. It's as if the virus can't make up its mind whether it wants to have an RNA genome or a DNA genome! But it gets even stranger.

Once completed, the reverse-transcribed DNA strand is then used as a template by the reverse transcriptase enzyme in an attempt to make a viral genome composed of double-stranded DNA. However, the viral enzyme must race to finish this second DNA strand before the viral capsid is completed around it—and capsid assembly always wins! The result is a DNA genome which is only partly double-stranded. No other virus replicates this way, so we can be pretty sure that hepatitis B virus arose from a unique evolutionary event.

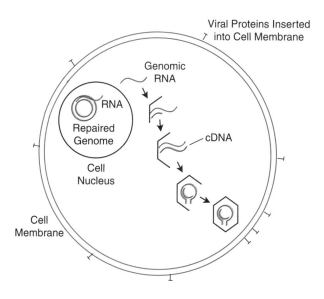

The hepatitis B genome encodes several proteins (for example, the major surface antigen, HBsAg) which are inserted into the membrane of the endoplasmic reticulum of the infected cell. There these proteins assemble to produce viral envelopes that bud into the interior of the endoplasmic reticulum. If a protein capsid containing viral DNA binds to the outside of the endoplasmic reticulum when one of these envelopes is assembling, it will be included inside the envelope, transported to the surface of the cell, and released into the surrounding tissues. However, when the virus makes these envelopes, it doesn't seem to care whether they contain completed capsids or not! In fact, a hepatitis B-infected liver cell produces about a thousand times as many empty viral envelopes as full. Production of this many non-infectious particles may appear wasteful, but as we shall see, these empty particles play a major role in the virus' ability to evade host defenses.

Another important feature of hepatitis B virus' reproductive style is that although infected liver cells can produce large quantities of virus, these cells usually are not killed by the viral infection. As a result, infected liver cells become factories that produce virus, yet continue to function relatively normally. In fact, hepatitis B-infected cells eventually produce so much virus that the blood of infected individuals (carriers) frequently contains about 100 million infectious particles per milliliter.

Viral Spread

Hepatitis B is spread efficiently by blood-to-blood contact, and in its most natural setting, this spread takes place during the birth of a child (so-called "vertical" transmission). I say "most natural" because, although hepatitis B virus can be efficiently spread, for example, when drug addicts share needles, the virus certainly did not evolve to be spread in this way. No, it is most likely that hepatitis B first learned to spread from human to human via the "perinatal" route—by passing from the blood of an infected mother to the blood of her child during childbirth. Indeed, about 20% of babies born to hepatitis B-infected mothers will be infected at birth.

There are several reasons why blood-to-blood spread of hepatitis B virus is so efficient. First, because of its reproductive strategy, infected cells become virus factories that continuously produce infectious virus. In addition, because of the structure of its lipoprotein envelope, hepatitis B virus is well suited to resist the assaults of enzymes that are found in the blood. As a result, large quantities of virus particles can accumulate over time in the blood of an infected individual. Finally, hepatitis B targets the liver—a large organ which is strategically positioned to intercept blood as it circulates through the body. Indeed, about 25% of the total cardiac output passes through the liver with each beat of the heart. This means that cells in the liver get a "good look" at any virus particles that have been introduced into the bloodstream. Because of these factors, hepatitis B virus ranks as one of the most infectious of all viruses: Transfer of a fraction of a drop of blood (as little as one microliter) is sufficient to spread the virus from one human to another.

Another rather natural route of hepatitis B transmission is from child to child, probably through open sores or cuts. This is most likely to be a preferred route when many children are crowded together, and hygiene is lax (e.g., in some daycare centers). In fact, because carriers of hepatitis B usually have so much infectious virus in their blood, any scenario you can imagine in which blood or blood products are exchanged has a high probability of transmitting a hepatitis B infection.

Hepatitis B virus is also found in seminal fluid. Although sexual transmission of hepatitis B virus is common, it is relatively inefficient. Indeed, studies show that mates of infected individuals frequently are not infected even after years of sexual contact.

Evasion of Host Defenses

Hepatitis B uses two main strategies to evade host defenses. First, the virus "gently" infects its target cells, usually without killing them. In fact, when a person is first infected, almost two months can elapse before significant amounts of virus are produced. Viruses like hepatitis B which do not kill their host cells (the "non-cytolytic" viruses) present a major problem for immune surveillance. This difficulty arises because before the adaptive system can be activated to make antibodies and killer T cells, the innate system must sense that there is danger. And one of the main clues that a dangerous viral attack is underway is the death of infected cells. Because hepatitis B usually does not kill the cells it infects, the innate system likely must wait for the odd infected liver cell to "make a fatal mistake" before it is alerted. Importantly, hepatitis B does not activate the interferon defense system, despite the fact that it is an enveloped virus. This suggests that it is not the presence of an envelope *per se* which triggers the interferon system, but rather the particular proteins (viral or cellular) that make up the viral envelope.

So by not killing the liver cells it infects, and by not inducing interferon production, hepatitis B virus can sneak up on the immune system. Eventually, however, virus-specific B cells are activated, and they begin to produce antibodies that recognize the major viral antigen on the surface of the viral envelope, HBsAg. However, even though they are present, antibodies to the HBsAg are not easily detected until rather late in an infection. This is because the empty viral envelopes also carry this antigen on their surfaces, and there are so many of them that they "soak up" the relatively smaller number of anti-HBsAg antibodies. It is only after the immune response has cleared the viral infection that anti-HBsAg antibodies are revealed. In fact, the appearance of anti-HBsAg antibodies is diagnostic of the fact that the immune system has "won" the battle with the virus.

The production of empty envelope "decoys," which soak up neutralizing antibodies, is the second strategy that hepatitis B virus uses to elude the immune response directed against it. The lack of effective neutralizing antibodies, which otherwise could bind to infectious viral particles and prevent them from infecting

liver cells, places the burden of repelling the attack squarely on the shoulders of virus-specific killer T cells. In many cases, killer T cells do eradicate the virus, but in other situations, the virus establishes a chronic, usually lifelong infection.

Pathogenesis

Hepatitis B virus is one of the world's most important pathogens, with about 500 million people carrying this virus as a chronic infection, and over a million dying each year of hepatitis B-associated liver disease. Based on our understanding of how hepatitis B virus solves its problems of reproduction, spread, and evasion, we can predict that the pathological consequences of a hepatitis B infection might occur according to the following scenario:

The infecting virus enters the bloodstream, and is carried to the liver. In liver cells, viral reproduction begins, and both empty and infectious virus particles are produced. As the newly minted viruses enter the blood and are recirculated to the liver, more liver cells are infected and these cells begin to produce even more virus particles. Sometime during these early weeks of infection, probably as a result of the death of a minority of the infected cells, the innate system senses that liver cells are being damaged, and initiates an inflammatory reaction in the liver. Cytokines produced by the innate system stimulate dendritic cells in the liver to migrate to nearby lymph nodes, carrying with them virus that has been acquired in the liver. Once in the lymph nodes, these antigen presenting cells activate virus-specific B and T cells. After a delay of a week or two, antibodies that recognize viral antigens begin to be produced, and killer T cells are activated that can recognize and kill liver cells infected by the virus. By this time, however, the virus has established a firm foothold in the liver, so the warriors of the immune system have their work cut out for them.

Despite its slow start, in about 70% of infected adults, the immune response to the infection is vigorous enough that the virus is eradicated with few or no symptoms. For the other 30%, direct destruction of liver cells by killer T cells, and indirect damage to liver cells due to the inflammatory reaction orchestrated by virus-specific helper T cells is severe enough to cause the symptoms commonly associated with liver damage: elevated levels of liver enzymes (e.g., aminotransferases) in the serum, nausea, vomiting, liver pain, jaundice, and dark-colored urine. These symptoms can last for several months while the immune system is battling to subdue the virus.

Fortunately, the immune system is usually victorious. In about 90% of symptomatic adults, the virus is eradicated, and damaged liver cells are replaced through the proliferation of healthy liver cells. However, in about 10% of symptomatic adults (roughly 3% of those initially infected), hepatitis virus wins the battle with the immune system, and establishes a chronic infection.

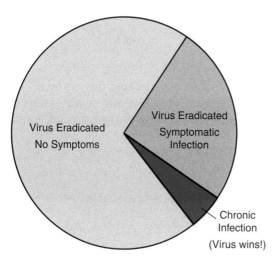

In newborns (the "natural" host for this virus), these numbers are very different: About 90% of infected newborns end up as chronic carriers of the virus. This difference presumably reflects the immature state of the newborn's immune system. The high proportion of infants infected at birth who go on to become chronic carriers eventually provides a relatively large pool of mothers who can infect their offspring, passing the infection from generation to generation.

Because neutralizing antibodies, which normally should mark viruses for destruction, are soaked up by empty envelope decoys, large amounts of infectious hepatitis B virus remain in the circulation of chronically infected individuals. As uninfected liver cells proliferate to replace those killed by the immune response, these "fresh" liver cells can be infected by the circulating virus particles, perpetuating the chronic infection.

One of the intriguing features of a chronic hepatitis B infection is that although most liver cells are infected, there is not mass destruction of the liver by killer T cells. Recent experiments suggest that cytokines produced by immune system cells can interfere with the production of hepatitis B virus, and in some cases can eliminate the virus from infected cells—without killing these cells. Exactly how this happens is not clear, but this might also explain why a chronic hepatitis B infection is generally characterized by periods during which the virus seems to be "hidden" from the immune system (perhaps because viral reproduction is suppressed), and periods during which new viruses are actively produced. Indeed, the impression one gets is that the im-

mune system warriors march into the liver and try to deal with the virus-infected cells–but can't quite get the job done. Then later, when the immune response has "relaxed," more virus is produced, and the cycle repeats.

Complexes between the abundant decoy viruses in the blood and the antibodies that bind to them can lead to skin rashes, painful joints, and kidney disease when decoy-antibody complexes are deposited in these areas. Hepatitis B virus usually does not kill the liver cells it infects, so most of the destruction of liver cells that occurs during a hepatitis B infection is the result of the immune response to the virus attack. Indeed, very little liver damage (cirrhosis) occurs in most patients whose immune systems are impaired.

For the virus, establishing a chronic infection in which high levels of virus persist in the blood makes perfect sense. The chronic infection allows hepatitis B virus to persist indefinitely in the infected host, and the plethora of infectious virus insures that a "carrier" will have a high probability of infecting her child at birth. However, for the human host, a chronic hepatitis B infection is certainly not advantageous. Although the immune systems of some chronically infected patients eventually eradicate the virus, most chronic infections are lifelong. Many of these patients remain relatively asymptomatic, whereas others suffer recurring bouts of liver inflammation which lead to more and more liver damage.

Another disease associated with chronic hepatitis B infections is liver cancer (hepatocellular carcinoma). Although there is no human virus which can be said to "cause" cancer, infection with some viruses can increase the risk that an individual will accumulate the mutations required to turn a normal cell into a cancer cell—and hepatitis B is an excellent example of such a virus. Indeed, roughly 20% of long-term, hepatitis B carriers eventually can be expected to contract liver cancer, and about one million people die of hepatitis B-associated hepatocellular carcinoma each year.

Exactly how this virus acts as a risk factor for liver cancer is not well understood, although it is likely that several facets of the virus' lifestyle play a role. First, one of the primary duties of liver cells is to detoxify potentially damaging chemicals that either enter the blood from outside the body, or which originate within the body as toxic byproducts of normal cellular metabolism. Many of these toxins (the "genotoxins") can directly or indirectly damage cellular DNA, and although liver cells are stocked with enzymes that can detoxify these chemicals, these detoxification systems can be overloaded. When this happens, liver cells become targets of the very genotoxins they normally are able to protect against.

Usually, liver cells are not proliferating, and in a resting state they frequently have time to repair damage inflicted by genotoxins. However, in a hepatitis B-infected liver, cells must proliferate to replace those that have been killed by the immune response to the infection. This "extra proliferation" increases the risk that these liver cells will divide before DNA damage can be repaired. So the combination of being constantly exposed to genotoxins while being forced to proliferate may predispose liver cells to cancer-causing mutations.

If exposure to genotoxins and increased proliferation of liver cells were all there was to the story, we would predict that many of the liver cells which eventually become cancerous would be uninfected, "bystander" cells that proliferated to replace infected cells killed by the immune response. However, experiments have shown that most hepatitis B-associated liver cancers contain remnants of hepatitis B genomes, often as DNA fragments that have somehow been inserted into the chromosomes of the cell. This means that most hepatitis B-associated cancer arises in cells that have been infected with the virus, and suggests that the virus does something to these cells which actively contributes to the cancer-causing process. What might this be?

Most research designed to discover which hepatitis B viral functions might be involved in cancer has centered on the viral "X" protein, aptly named because its role in hepatitis B infections remains mysterious. Some experiments indicate that the X protein can bind to and alter the function of the p53 tumor suppressor protein. When a cell with unrepaired DNA damage begins to proliferate, that cell is usually stopped in its tracks by the action of the p53 tumor suppressor—the so-called "guardian of the genome." If the DNA damage is minor and can be repaired, proliferation is halted until the repair has been completed. On the other hand, if DNA damage is extensive, the p53 protein triggers the cell to commit suicide. So in hepatitis B-infected cells, the normal safeguard against DNA damage may not be in place, because the X protein may inactivate the p53 guardian function. This would make cells infected with hepatitis B more susceptible to cancer-causing mutations.

I say "may inactivate p53 function," because this issue is still controversial. Most researchers believe that the viral X protein probably plays some role in hepatitis B virus-associated liver cancer, but there is disagreement on just what that role might be. I should point out that although virologists aren't clear about how the X protein works, it is known that the X protein is essential for viral reproduction and spread within infected hosts. So hepatitis B didn't "invent" this protein just to increase cancer risks.

By choosing to infect liver cells that constantly are subjected to DNA-damaging toxins, by causing increased liver cell proliferation, and perhaps by disabling the p53 tumor suppressor, hepatitis B virus increases the probability that liver cells will collect the mutations required for tumorigenesis. Still, it usually takes many years for these mutations to accumulate, and hepatitis B-associated liver tumors generally arise twenty to fifty years post infection.

HEPATITIS C VIRUS— AN ESCAPE ARTIST

Hepatitis C is the third virus in our Bug Parade that causes liver damage. This virus was first identified in 1989, but it isn't clear whether hepatitis C is really a "new" virus, or whether advances in technology have only recently made it's detection possible. So far, it has been very difficult to grow this virus in the lab, making it hard to learn its secrets. Nevertheless, hepatitis C virus now infects about 170 million people worldwide, and many of these people will suffer from cirrhosis of the liver or liver cancer. So even though we are working with inadequate information, we need to do our best to understand this virus.

Viral Reproduction

Hepatitis C virus infects both liver cells and macrophages. Recent experiments suggest that the virus binds to low density lipoproteins in the blood, and then enters its target cell when these virus-lipoprotein complexes bind to receptors on the cell's surface. Of course, every living cell in the body has receptors for low density lipoproteins—that's how they get the cholesterol required to repair or synthesize cell membranes. Although the liver has relatively more of these receptors than most cells, the intracellular environment of liver cells and macrophages must also play a major role in target cell selection. The lack of a good system for growing hepatitis C in cultured human cells has made it impossible to determine whether this virus kills the cells it infects, so a critical piece of information for understanding the lifestyle of this virus is missing.

Estimates of the fraction of liver cells infected during a chronic hepatitis C infection vary wildly from less than 5% to 100%, but it is known that about one trillion virus particles are produced each day. Because there are roughly 200 billion liver cells (hepatocytes) in an adult human, each liver cell would only have to produce about 100 viruses per day, even if as few as 5% of them were infected. Clearly, if a virus infects a big organ (like the liver), only a modest amount of viral reproduction is required per cell to produce a huge number of viruses.

The hepatitis C virus genome consists of a single strand of positive-sense RNA enclosed in an envelope acquired from the infected cell. Viral RNA is translated in a cap-independent fashion (similar to rhinovirus) to yield a single, long polyprotein. This monster protein is subsequently cleaved at specific sites to yield the functional viral proteins. One important feature of hepatitis C replication is that new hepatitis C genomes are produced by a virus-encoded RNA polymerase that is extremely error-prone.

Spread of Hepatitis C Virus

In the United States, it is estimated that about 2% of the population is infected with hepatitis C. This means that roughly five times as many Americans are infected with this virus as with HIV-1. Transmission of hepatitis C by "unnatural" routes involving the transfer of blood (e.g., by transfusions or by intravenous drug abuse) or blood products (e.g., clotting factors) is very efficient. This makes sense: A large number of hepatitis C viruses are found in the blood of infected individuals; this virus has a protective envelope that resists attacks by the complement system, which would destroy less well-protected viruses in the blood; and hepatitis C virus infects the liver, through which large volumes of blood circulate continuously.

In contrast, the natural routes of infection available to this virus have proved more difficult to elucidate. Although the virus is found in saliva and semen, transmission of hepatitis C virus by sexual contact appears to be inefficient. Indeed, the only well-documented "natural" avenue of transmission is from infected mothers to their offspring when blood of the mother is introduced into the bloodstream of the newborn as a result of trauma during childbirth. However, only about 7% of children born to infected mothers are infected by this route, so there must be more to the story. In addition, roughly half of all hepatitis C patients claim not to have been exposed to blood or blood products, so it is likely that there are other routes of infection which have not been discovered—or that a lot of people have "forgotten" that they received a blood transfusion or that they injected drugs. So far, humans appear to be its only natural host, so there is a lot of mystery surrounding the spread of hepatitis C virus.

Viral Evasion

As with any virus that is transmitted from mother to child, hepatitis C virus must find a way to persist in the infected host until she is old enough to pass the virus along to her offspring. Hepatitis C accomplishes this feat by establishing a life-long, chronic infection. Indeed, roughly half of all individuals infected with this virus become chronically infected "carriers." How hepatitis C virus manages to evade the immune system during this chronic infection is not well understood—but there are some clues.

When hepatitis C viruses are isolated from a single infected carrier and the genomes of these isolates are sequenced, a large number of genetically different, but closely related, "quasispecies" are found. This indicates that in hepatitis C infections, the virus' error-prone RNA polymerase is fully utilized to produce a high rate of mutation in the viral genome. In addition, when aminotransferase levels (an indicator of liver damage) are measured in chronically infected individuals, they generally fluctuate like a sine wave with a period of roughly six weeks. Taken together, these observations suggest that the virus' high mutation rate allows it to "escape" from the host's immune defenses. During the initial infection, antibodies and killer T cells are produced that recognize the infecting virus. Then, just when the immune system has almost wiped out the virus, mutations in the viral genome give rise to one or more new variants which cannot be recognized by the killer T cells or antibodies. As a result, the immune system must start from scratch to produce new B cells and killer T cells which are appropriate to defend against the mutated virus. Then, just as these new B and T cells go to work, new mutations allow the virus to escape again. It is likely that this cycle of killing and escape plays a major role in the ability of hepatitis C virus to maintain a chronic infection.

When you think about it, this type of evasion must be managed carefully. If the virus mutates too slowly, the immune system will "win" and the virus will be eradicated before it can spread. On the other hand, if the mutation rate is too great, the immune system will be completely ineffective, the virus will reproduce without restraints, and the host will probably be killed. So it's likely that during viral evolution, it took some "tuning" to perfect this evasion strategy. The fact that hepatitis C virus establishes a chronic infection in the great majority of infected humans, and that these carriers usually live for many decades without serious symptoms is proof that hepatitis C did eventually get it right.

Although the antibodies and killer T cells of the adaptive immune system play critical roles in controlling viral spread, a virus that wishes to establish a chronic infection must also figure out how to evade the <u>innate</u> immune system. Hepatitis C has a lipid envelope and replication of its genome involves a double-stranded RNA intermediate. Consequently, this virus can be expected to activate the interferon defense system. This defense is so potent, that without a good way of evading it, no virus could establish a chronic infection. Recent research suggests that at least two hepatitis C proteins (E2 and NS5A) may help protect virus-infected cells against the effects of interferon, probably by inhibiting the function of PKR. Interestingly, these interferon evasion tactics appear to be more effective in certain hepatitis C genotypes. This has the practical implication that some hepatitis C genotypes are more amenable to treatment with interferon than are others.

Pathological Consequences

Our hypothesis is that if we can understand a virus' lifestyle, we should then be able to predict what diseases the viral infection will cause. However, in the case of hepatitis C, we still don't have a full picture of how the virus reproduces, spreads, and evades host defenses—so predicting the pathological consequences of a hepatitis C infection is a little "iffy." This situation illustrates how important basic research in virology is to understanding viral disease. It is only when you can "fathom the mind of a virus" that you can understand how it impacts its host, and can rationally design therapies that might lessen the effects of a viral infection or that might help host defenses eliminate the virus. Fortunately, research on hepatitis C virus is now proceeding at a rapid rate, so some of the remaining questions about this virus' lifestyle may be answered by the time you see this lecture. For now, I will sketch a likely scenario to explain the pathology, based on what virologists have discovered so far.

At least two-thirds of all new hepatitis C infections go unrecognized because they are asymptomatic or only mildly symptomatic. In fact, hepatitis C usually establishes a smoldering, chronic infection that is not detected until ten or more years after the initial exposure. The fact that hepatitis C infections frequently are asymptomatic for so long suggests that the virus reproduces slowly in infected liver cells, and that the immune system is effective at controlling the infection—but not quite effective enough.

To maintain a chronic infection, hepatitis C uses its error-prone polymerase to generate mutations which keep the virus one step ahead of host defenses. In about 20% of chronic infections, the continuing waves of infection that occur as mutant viruses escape from immune surveillance eventually lead to cirrhosis of the liver. This generally occurs about two decades post infection, and results when the killing of liver cells (either by the virus or by the immune response to the infection—or both) produces scar tissue which disrupts the architecture and function of the liver. It is interesting, but not unexpected, that hepatitis C infections and excessive alcohol consumption can act synergistically to accelerate the progression from infection to cirrhosis.

About 10% of patients with hepatitis C-induced cirrhosis of the liver eventually suffer from liver cancer (hepatocellular carcinoma) that usually arises about three decades post infection. How the hepatitis C infection predisposes a person to liver cancer is not known for certain. It may be that the proliferation of liver cells, which is required to replace cells destroyed by the infection, increases the probability that mistakes will be made during cellular DNA replication and that the replacement cells will become cancerous. Also, there are reports that one of the hepatitis C proteins (NS5A) can associate with the tumor suppressor protein, p53, and interfere with its function. So it may be that proteins required for the virus life cycle may also (inadvertently) play a role in hepatitis C-associated liver cancer. Clearly, more basic research needs to be done before a firm connection between the virus lifestyle and liver cancer can be established.

A Comparison of Three Hepatitis Viruses

Now that we have discussed three different viruses that cause hepatitis—hepatitis A, B, and C viruses—it's appropriate to consider the similarities and differences in the way these three viruses do business. Hepatitis A and C viruses have RNA genomes whereas hepatitis B virus' genome is made of DNA. Hepatitis B and C are enveloped viruses, but hepatitis A makes do with one really tough protein capsid. Hepatitis B and C are spread mainly by the transfer of blood, but hepatitis A virus is usually spread by the fecal-oral route. All three viruses target liver cells for infection, but hepatitis A first infects intestinal cells, and then takes a detour through the liver on the way back to the intestinal tract.

Hepatitis A virus usually does very little liver damage, and never establishes a chronic infection. As a result, although about half of all Americans have been infected with this virus, almost all have healthy livers and are virus free. In contrast, both hepatitis B and C viruses efficiently establish chronic infections. They have to do this, because their "natural" route of infection is from mother to child during childbirth—a route which becomes a dead end if the virus does not persist in the infected child until she herself can become a mother. To make this work, hepatitis B establishes a lifelong chronic infection in about 90% of infected newborns, and hepatitis C appears to be similarly efficient in establishing chronic infections. Because both viruses can be spread, albeit inefficiently, by sexual intercourse, a chronic infection also serves to increase the probability that hepatitis B and C viruses will be spread during a lifetime of sex. Approximately 5% and 2% of Americans have been infected with hepatitis B and C viruses, respectively.

To evade immune defenses during its chronic infection, hepatitis B produces a huge number of decoy viruses that soak up antibodies which normally would tag the virus for destruction. As a result, infectious viruses circulating in the blood can infect susceptible liver cells as they become available, perpetuating the infection. In contrast, hepatitis C virus takes full advantage of its error-prone RNA polymerase to produce mutated viruses. These mutants can stay one step ahead of the immune defenses, allowing hepatitis C to persist in its infected hosts. The fact that these two viruses use very different strategies to establish lifelong infections has a practical implication. Because hepatitis B virus has essentially only one serotype, an excellent vaccine exists which can protect against hepatitis B infections. In contrast, a vaccine against hepatitis C is difficult to imagine because of the large number of quasispecies that exist as a result of this virus' high mutation rate.

During chronic infections with hepatitis B or C viruses, liver cells are destroyed as the host's immune system tries to rid the patient of the virus. Over the years, this destruction can lead to cirrhosis of the liver, and the chronic infection can also predispose the infected individual to liver cancer. In contrast, infection with hepatitis A virus, which does not cause a chronic infection, does not increase a person's chances of contracting liver cancer. This underscores the concept that only viruses that cause long-term infections are associated with cancer in humans.

HUMAN T CELL LYMPHOTROPIC VIRUS TYPE I—A TRIBAL VIRUS

Retroviruses, which have RNA genomes, but which replicate through a DNA intermediate, were first isolated from animals at the beginning of the nineteenth century. However, it was not until 1980 that the first human retrovirus, human T cell lymphotropic virus type I (HTLV-I) was discovered. Since then, other human retroviruses have been identified, and HTLV-I (a.k.a. human T cell leukemia virus) has been overshadowed by its more famous cousin, the human immunodeficiency virus (HIV-1) which causes AIDS (originally called HTLV-III).

Viral Reproduction

HTLV-I enters its target cell when its envelope binds to an unknown receptor on the cell's surface, and then fuses with the cell membrane. This "injects" the viral capsid, which encloses two copies of HTLV-1's single-stranded RNA genome, into the cytoplasm of the cell. There the capsid is removed, and a viral enzyme (reverse transcriptase), which is packaged in the capsid, springs into action. This enzyme copies the RNA genome to produce a single-stranded, complementary DNA (cDNA) molecule, destroying the original RNA molecule after it has been copied. The reverse transcriptase protein then makes a complementary copy of the single cDNA strand to produce a double-stranded cDNA molecule. The net result of all this action is to replace the single-stranded RNA genome with a double-stranded DNA "copy" that contains the viral genetic information.

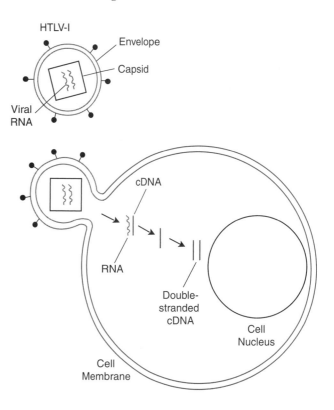

This double-stranded copy of the original viral RNA then enters the cell's nucleus where a viral enzyme (integrase) cuts the cellular DNA, and inserts the double-stranded viral DNA at this spot.

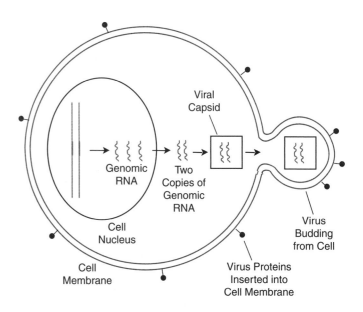

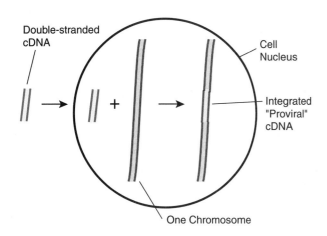

Although the chromosomal site where the cDNA is inserted is essentially random, the viral cDNA is carefully integrated so that none of the viral genes is interrupted. HTLV-I's ability to actively integrate genetic information for the complete virus into the target cell's DNA is key to producing a lifelong infection: Once integrated into the cellular genome, the HTLV-I sequences will be copied and passed down to daughter cells as the infected cell replicates its DNA and divides.

From its position within one of the cell's chromosomes, the integrated viral genome can be transcribed by the cell's RNA polymerase to make either short mRNAs, which encode the various viral proteins, or full-length RNAs that will be used as new viral genomes. Eventually, two of these genomic RNAs are packaged into each protein capsid, and the virus leaves the cell by budding through the cell membrane, picking up an envelope which contains both viral and cellular proteins. Importantly, HTLV-I-infected cells usually are not killed in the process of producing new viruses.

Another important feature of this virus' lifestyle is that no HTLV-I mRNAs or genomes can be produced until the double-stranded copy of the viral genome has been inserted into the cellular genome. Moreover, this integration event cannot take place unless the target cell is in the process of replicating its cellular DNA. This means that resting cells (most of the cells in the body) are not good targets for HTLV-I infection.

The number of viruses produced in an HTLV-I infection is very small. In fact, once the double-stranded DNA copy of the virus (the "provirus") has been integrated into a cell's genome, it usually just sits there. In this silent (latent) infection, little or no viral mRNA is made, and no new virus particles are produced. Occasionally, the proviruses in latently infected cells "reactivate" and begin to produce new virus particles. How HTLV-I manages to remain silent for so long in its latent state, and how it is reactivated are unsolved mysteries. This is not unexpected: There is not a single virus for which the process of latency is completely understood. In any case, the typical result of an HTLV-I infection is that the infected individual becomes a carrier of the virus for life, with the virus persisting within infected cells in a latent state.

Virus Spread

HTLV-I primarily infects T cells which express CD4 proteins on their surfaces. These are mainly helper T cells, and it is not known whether the preference for these cells is a consequence of the virus binding to surface proteins that are peculiar to $CD4^+$ cells, or whether these cells simply present an environment that is permissive for the virus' life cycle.

This brings us to an interesting conundrum. People infected with HTLV-I generally have many $CD4^+$ T cells that have the provirus integrated into their genomes. In fact, in established HTLV-I carriers, roughly 1% of their $CD4^+$ T cells (about ten billion) have an integrated provirus. Yet despite this huge number of infected cells, HTLV-I carriers usually have almost undetectable amounts of virus circulating in their blood. In addition, experiments have shown that HTLV-I is not very infectious—even for its favorite target, the $CD4^+$ T cell. So how can a virus which has a hard time infecting its target cell, and which doesn't produce many virus progeny even when an infection is successful—how can such a virus possibly infect 1% of a carrier's $CD4^+$ T cells? Moreover, how can such a wimpy virus ever hope to spread efficiently enough from human to human to survive? As you will soon see, this "wimpy virus" is very clever indeed, having solved both these problems in elegant ways.

After the HTLV-I provirus has been integrated into the cellular genome, transcription of viral genes can begin. In addition to genes that encode the viral reverse transcriptase and the viral coat proteins, the HTLV-I provirus also includes a gene for a protein called Tax. This protein is produced early after infection, and functions as a transcription factor that can cause HTLV-I-infected cells to proliferate by modulating the expression or function of cellular genes. When these infected cells proliferate, the integrated provirus is copied right along with the rest of the cell's DNA, and is passed down to each of the daughter cells, resulting in two infected cells. Next, the Tax protein produced in these daughter cells causes them to proliferate, resulting in four infected cells. And so it continues. Consequently, by first integrating its genetic information in the form of a provirus into the genome of a $CD4^+$ T cell, and then by driving this cell to proliferate, HTLV-I is able to "infect" a large fraction of a person's $CD4^+$ T cells without needing to produce infectious virus.

So both hepatitis B virus and HTLV-I can establish lifelong infections, but the mechanisms they use to accomplish this are very different. Whereas hepatitis B infections are probably maintained by viruses that are circulating in the blood, HTLV-I is "immortalized" by inserting cDNA copies of its RNA genome into the chromosomes of the cells it infects. Then, when these infected cells proliferate, the HTLV-I infection is amplified and perpetuated.

Although the proliferation of HTLV-I-infected cells likely accounts for most of the spread of HTLV-I within an infected individual, there are two other ways that the number of infected cells can be increased. First, HTLV-I can spread when infected $CD4^+$ cells fuse with uninfected $CD4^+$ cells to produce giant cells (syncytia) that contain the provirus. This then is a second way HTLV-I can infect cells without producing any new virus particles. In addition to these two "virus-less" infection strategies, viruses produced by infected cells also can infect other $CD4^+$ T cells, albeit inefficiently. In fact, to try to make infection by virus particles more likely, HTLV-I-infected cells can trigger proliferation in nearby, uninfected T cells. Exactly how this "bystander activation" works is being debated. One possibility is that the viral Tax protein increases expression of certain proteins on the surface of the infected cell (e.g., adhesion proteins) which then contact proteins on the surface of an adjacent, uninfected $CD4^+$ cell, causing it to proliferate. Alternatively, the Tax protein may induce the infected cell to produce and export growth factors (e.g., interleukin-2) that can signal other T cells in the neighborhood to proliferate. Whatever the mechanism, inducing other $CD4^+$ cells to proliferate is in the virus' best interest, since during infection, proviral DNA can only be integrated into the genomes of proliferating cells.

Clues as to how HTLV-I spreads from human to human come from an analysis of the incidence of HTLV-I infections in persons who received blood or blood products from HTLV-I-infected blood donors. These studies show that hemophiliacs are not at risk of contracting HTLV-I from clotting factor VIII prepared from cell-free blood donated by HTLV-I-infected donors. In contrast, recipients of whole blood transfusions from HTLV-I-infected donors frequently become infected. These and other observations indicate that HTLV-I infections spread most efficiently when HTLV-I-infected cells, not virus particles, are transferred from one human to another. There are two likely scenarios that could explain how an HTLV-I infection spreads by the transfer of infected cells. First, transferred infected cells may subsequently produce virus particles which can then infect $CD4^+$ T cells of the recipient. Alternatively, infected "donor" $CD4^+$ T cells may fuse with uninfected $CD4^+$ T cells of the recipient. These "double cells" contain the integrated provirus contributed by the infected cell, so they are capable of producing virus particles that can then infect other cells of the new host.

Infected cells are present in blood, in breast milk, in semen, and to a much lesser extent, in vaginal secretions of persons who have been infected with HTLV-I. Importantly, the likelihood of an infected mother spreading the infection to her baby by breast feeding has been reported to be as high as 38%, with the probability of spread increasing with longer periods of breast feeding. The virus-infected cells which these babies receive with mom's breast milk can establish a persistent infection that lasts for life. Consequently, once these infected babies reach puberty, they can infect their mates during sexual intercourse or pass the virus to their offspring during breast feeding. Overall, however, HTLV-I is not very infectious, so spread requires either a lot of breast feeding or a lot of sex. Nevertheless, by establishing a latent infection, HTLV-I makes up for its low infectivity by giving an infected person a whole lifetime of chances to pass it on.

One of the fascinating features of HTLV-I infections is that they occur most frequently in geographically isolated populations (e.g., small villages on Japanese islands). Further, when people from these areas move to larger population centers, they remain susceptible to infection for a while, and then, with time, their rates of HTLV-I infection decrease. Although virologists are still not sure why high rates of HTLV-I infection are found in isolated populations, it is possible that there is something in the genetic makeup of people in these small "tribes" which makes it easier for the virus to spread among them. This would make sense, because HTLV-I is believed to be a very old virus that "jumped" from a nonhuman primate into the human population roughly 100,000 years ago—at a time when most humans lived in small, isolated tribes.

One possibility for the tribal nature of HTLV-I infections is that members of these tribes just happen to have more or better receptors for HTLV-I on the surfaces of their T cells, making them stronger candidates for a viral infection. Unfortunately, because the cellular receptors for HTLV-I have not been identified, this hypothesis cannot be tested. Alternatively, the virus may have jumped from a nonhuman primate into a tribe, and once seeded, the virus may have spread within this relatively closed population until a large fraction of the tribe became infected. Later, the virus may have been transmitted by infected tribe members to other, nearby tribes.

There is another hypothesis, however, which I think is even more interesting, and that could explain why HTLV-I spreads best in old, isolated populations.

When you think about it, transferring a cell, for example an HTLV-I-infected cell, from one human to another is an act of "transplantation." When transplant surgeons transfer an organ from one person to another, they try to use donors and recipients whose MHC molecules are as similar as possible. The reason for this is that unmatched MHC molecules, which appear on the surfaces of cells in the transplanted organ but not on the surfaces of the cells of the recipient, are favorite targets for attack by the recipient's killer T cells. And the poorer the MHC match, the more vigorous will be the attack. Indeed, most transplant recipients must be immunosuppressed for life (e.g., by the administration of cyclosporin) to prevent destruction of the donated organ by the recipient's killer T cells—destruction which can begin within minutes after transplantation.

Now if you live in a large country like the United States, which is a "mixing pot" for people of many different genetic backgrounds, there are so many different varieties of MHC molecules in this "outbred" population that it is extremely difficult to find an organ donor whose MHC molecules are a close match to your own. In contrast, if you live on an isolated island with a very small population, it's quite likely that you will end up marrying your cousin, or your half-sister, or even your sister. There just aren't that many people of an appropriate age to choose among. If this type of intermarrying goes on long enough, the genetic makeup (including the MHC genes) of your tribe will become more and more similar, and the chances of finding a member of your tribe with a perfect or near-perfect match of MHC molecules will be greatly increased. So if you need an organ transplant, it would be helpful to live in a small, isolated tribe that has existed for a long time.

Likewise, a virus like HTLV-I, which spreads more easily by the transfer of infected cells than by the transfer of virus, might also prefer to "live" in an isolated, inbred tribe. When HTLV-I-infected cells are transferred, these cells become targets for killer T cell attack if their MHC molecules differ from those of the recipient. In an inbred population, there are relatively fewer different MHC molecules, so the probability of matching MHC molecules would be better. As a result, when HTLV-I-infected cells are transferred (e.g., during sex), they should have a higher probability of surviving long enough to produce virus before being destroyed by the recipient's killer T cells.

If HTLV-I evolved to thrive in small, isolated populations, you might predict that as these tribes are mixed into the general population, HTLV-I might be destined for extinction. In fact, epidemiological studies (e.g., in Japan) indicate that the number of persons newly infected with HTLV-I is decreasing. Part of this decrease is certainly due to recent programs that discourage breast feeding. However, the decline in new HTLV-I infections began before anti-breast feeding programs were instituted, and before the blood supply could be tested for this virus.

So it could be argued that the cases of HTLV-I infection we are seeing now are "leftovers" from a virus that evolved to spread in what we might call tribes. This is in contrast to hepatitis B virus which, because it is so infectious, plagues more and more people worldwide each year. However, two new susceptible populations have emerged which may "rescue" HTLV-I from extinction: intravenous drug abusers and persons infected with the AIDS virus (HIV-1). Intravenous drug abusers can spread HTLV-I-infected cells very efficiently by sharing contaminated needles, and many persons infected with HIV-1 engage in unprotected, promiscuous sex, increasing the number of people that a single infected individual is likely to infect. In addition, members of both groups tend to be immunosuppressed, either by virtue of an HIV-1 infection or because of a generally unhealthy lifestyle. And if the "MHC barrier" hypothesis is correct, immunosuppression of a potential infectee should make it more likely that transferred HTLV-I-infected cells would survive immune rejection.

Viral Evasion Tactics

Clearly, the most effective evasion tactic employed by HTLV-I is its ability to establish a latent infection in which the integrated provirus is passed down to daughter cells when infected cells proliferate. During a latent infection, little or no viral RNA is transcribed from the integrated, proviral genome. So to the immune system, latently infected cells are indistinguishable from uninfected cells. This is a good thing for the virus, because studies have shown that persons infected with HTLV-I produce large quantities of killer T cells that can recognize the viral Tax protein if it is presented by MHC molecules on infected cells. The choice of the Tax protein as a target is a smart one, because the Tax protein is produced early after infection, giving killer T cells a good chance to destroy infected cells before they can make more virus. The fact that the immune system is ready to pounce on latently infected cells when they reactivate probably explains why there is so little virus produced in persons infected with HTLV-I. This also suggests that during an HTLV-I infection, the virus must tread a fine line between producing Tax proteins which drive cell

proliferation (thereby amplifying the number of infected cells), and hiding from killer T cells that can destroy Tax-producing cells. I wish I could tell you how the virus manages to do this, but it's a mystery.

Because HTLV-I is an RNA virus which replicates using an error-prone reverse transcriptase, it might be expected that antigenic drift would be another of HTLV-I's evasion mechanisms. However, when proviruses isolated from cells of an infected person have been sequenced, they turn out to be amazingly homogeneous. This is probably because the virus spreads within an infected individual mostly by cell proliferation, not by repeated rounds of viral infection. During cell proliferation, the provirus is copied by the high-fidelity DNA replication machinery of the cell. In contrast, during rounds of viral infection, the error-prone viral reverse transcriptase is used to create DNA proviruses.

Viral Pathogenesis

Worldwide, about 15 million people are currently infected with HTLV-I, yet greater than 95% of these HTLV-I infections will never result in disease. In fact, the initial infection with HTLV-I is uniformly asymptomatic. This is to be expected since HTLV-I does not kill the cells it infects, and the virus establishes an infection in which the virus' genetic information is "silent" in an overwhelming majority of infected cells.

HTLV-I infections are associated with a human cancer—adult T cell leukemia (ATL). Fortunately, only about 2% of HTLV-I-infected individuals contract this form of leukemia. This is in contrast to hepatitis B virus carriers who have about a 20% risk of contracting liver cancer. As with hepatitis B-associated liver cancer, most cases of HTLV-I-associated leukemia do not develop until thirty to fifty years post infection. Clearly, leukemia is an unintended consequence of an HTLV-I infection: In the ancient population which this virus evolved to infect, people simply didn't live long enough to be at risk for virus-associated leukemia.

What feature(s) of an HTLV-I infection increases the probability that a chronically-infected person will develop leukemia? Cancer (e.g., leukemia) is believed to result when a cell accrues mutations that activate proto-oncogenes and disable tumor suppressor genes. Therefore, to understand how an HTLV-I infection could predispose a person to adult T cell leukemia, we really need to answer the question, "How does an HTLV-I infection increase the mutation rate in $CD4^+$ T cells?" The answer to this question is not known for certain, but there are several likely hypotheses which, not surprisingly, are based on what is known about how the virus solves its problems of reproduction, spread, and evasion.

To facilitate its spread, HTLV-I produces the Tax protein that can trigger both infected cells and uninfected, bystander cells to proliferate. This "extra" cell proliferation increases the risk that infected or bystander cells will accumulate mutations. Indeed, in chronically infected individuals, leukemic cells arise both from HTLV-I-infected cells and from bystander cells which have no integrated provirus. Moreover, as part of its strategy for triggering cell proliferation, HTLV-I uses its Tax protein to functionally disable the p53 "guardian of the genome," making infected cells even more susceptible to damaging mutations. So by inducing chronic cell proliferation and by depriving infected cells of the protection of the p53 tumor suppressor protein, HTLV-I likely increases the probability that a $CD4^+$ T cell in an infected individual will eventually accumulate the mutations required to turn that cell into a leukemic cell.

In addition to an increased risk for contracting adult T cell leukemia, persons infected with HTLV-I also have about a 2% lifetime risk of suffering from a neurological disease with a very long name: HTLV-I-associated, myelopathy/tropical spastic paraparesis (usually called HAM/TSP). HAM/TSP is a progressive, paralytic disease in which the long motor neurons in the spinal cord become demyelinated. Like adult T cell leukemia, HAM/TSP is usually seen thirty to fifty years post infection. Although the details are not clear, it is believed that HAM/TSP is caused by a dysregulation of the immune system precipitated by the stimulation of T cells over a period of many years in response to the chronic HTLV-I infection.

Table 5.1 compares and contrasts the lifestyles of the three viruses we discussed in this lecture.

Table 5.1 Viruses We Get from Mom

	Hepatitis B Virus	Hepatitis C Virus	HTLV-I
R E P R O D U C E	DNA genome is mostly double-stranded	Positive, single-stranded RNA genome	Single-stranded RNA genome
	Has protein capsid enclosed in cell-derived envelope	Enveloped virus	Has protein capsid enclosed in cell-derived envelope
	Bizarre replication strategy using reverse transcriptase enzyme—many empty particles generated	Error-prone RNA polymerase	Retrovirus—reverse transcriptase produces double–stranded DNA copy of genome that is integrated into host chromosome
	Non-cytolytic	Not known whether cytolytic	Non-cytolytic
S P R E A D	Infects liver cells (hepatocytes)	Infects liver cells and macrophages	Infects $CD4^+$ T cells
			Spreads within host mainly when infected cells proliferate
	Spreads to new hosts extremely efficiently by transfer of virus in blood	Spreads to new hosts by transfer of virus in blood, and by other, unknown routes	Spreads to new hosts by transfer of virus-infected cells in mother's milk
	Causes acute or chronic infection	Usually causes chronic infection	Causes chronic infection
E V A D E	Does not induce interferon production	Viral proteins protect against effects of interferon	Can establish stealth infection of $CD4^+$ T cells
	Produces decoy viruses that confuse antibody response	Mutates rapidly to "escape" immune response	

Viruses We Get by Intimate Physical Contact

R E V I E W

Let's review some important features of the three viruses we discussed in the last lecture: hepatitis B, hepatitis C, and HTLV-I. All three viruses are spread from mother to child—hepatitis B and C mainly by transfer, during the trauma of childbirth, of blood that contains infectious virus; and HTLV-I chiefly by transfer of virus-infected cells during breast feeding. Later, when infected children reach puberty, these viral infections can be spread during sexual intercourse, albeit inefficiently. To make mother to child transfer and sexual intercourse viable routes for viral spread, all three viruses can establish lifelong infections. However, the strategies these viruses use to persist in their infected hosts are very different.

HTLV-I is a retrovirus which employs the viral reverse transcriptase enzyme to produce a double-stranded DNA "copy" of the viral genetic information (the provirus). Another viral enzyme, the viral integrase, then carefully inserts the provirus into the genome of an infected $CD4^+$ T cell. Hepatitis B also encodes a reverse transcriptase that it uses to produce new viral genomes. However, hepatitis B lacks an integrase enzyme, so hepatitis B DNA is not integrated into the host cell's chromosome during viral replication. Once integrated, the HTLV-I proviral DNA can persist for the life of the host. This is a true "latent" infection in which most infected cells do not produce virus. The HTLV-I-encoded Tax protein induces both infected cells and uninfected, bystander T cells to proliferate—and when infected cells proliferate, the integrated provirus is copied and passed down to the daughter cells, increasing the number of cells that are infected. The induction of bystander cell proliferation is important, because some latently infected cells can reactivate to produce infectious virus—and these newly made viruses can infect only cells which are proliferating. Because HTLV-I does not kill the cells it infects, and because most infected cells produce little or no virus, initial HTLV-I infections are invariably asymptomatic.

Persons with chronic hepatitis B infections generally have high levels of infectious virus in their blood. These viruses escape eradication by the host's immune system because antibodies that could tag the viruses for destruction are soaked up by empty, decoy viruses which infected liver cells produce in abundance. As these viruses circulate through the liver with the blood, susceptible liver cells are continuously infected. So whereas HTLV-I persists in a latent state, hiding its genetic information in the chromosomes of infected cells, hepatitis B virus persists by maintaining a chronic infection in which there is a standoff between the immune system and the viral infection. Because hepatitis B virus also does not kill the cells it infects, these infections, too, are frequently asymptomatic, and when moderate to severe liver damage does occur, it is the result of the immune system's response to the hepatitis B infection.

Hepatitis C virus also persists as a chronic infection but the trick it uses to achieve this end is quite different. Hepatitis C virus uses its error-prone RNA polymerase to introduce mutations into newly made genomes, making it possible for the virus to stay one step ahead of the host's immune system. During a chronic hepatitis C infection, the immune system attempts to eradicate the virus. However, just when it looks like the immune response will win, the virus "escapes" by mutating to a new form which existing antibodies and killer T cells cannot recognize. Then, while the immune system is producing new antibodies and killer T cells, the mutant virus is free to infect additional liver cells. These periods of immune killing and virus escape alternate during the lifetime of the infected person, frequently leading to cirrhosis of the liver.

Because the blood of hepatitis B carriers is chock full of virus, and because any blood transferred to another human almost immediately comes in contact with liver cells, hepatitis B is one of the most infectious viruses known. In contrast, HTLV-I is not very infectious, relying mainly on the transfer of infected <u>cells</u> to spread from human to human. It is likely that these differences in infectivity and mode of infection (virus vs. virus-infected cells) explain why hepatitis B virus has spread easily all over the world, whereas HTLV-I has mainly infected members of isolated tribes and their descendants. The spread of hepatitis C virus is shrouded in mystery. Although the virus can be transmitted during childbirth, probably during sexual intercourse, and certainly during the exchange of blood or blood products, about half of all hepatitis C infections appear not to involve any of these routes.

Persons infected with HTLV-I have about a 3% lifetime risk of contracting adult T cell leukemia. Likewise, infection with hepatitis B or hepatitis C virus increases the risk for liver cancer. As is typical of virus-associated cancers, these malignancies usually arise decades post infection. It is presumed that during a lifelong infection, the virus somehow increases the rate at which infected cells or bystander cells accumulate mutations. HTLV-I and hepatitis B virus make proteins that are required for their reproduction, but which also are likely to be involved in the predisposition to cancer. Hepatitis C virus probably also encodes proteins that play a role in hepatitis C-associated liver cancer, but more research is needed to confirm this finding.

An additional pathological consequence of some HTLV-I infections is myelopathy/tropical spastic paraparesis (HAM/TSP), a progressive paralytic disease that arises in about 1% of infected individuals. Again, HAM/TSP usually occurs decades after infection, and probably results when T cells, which are trying in vain to eradicate the HTLV-I infection, are chronically stimulated.

VIRUSES WE GET BY INTIMATE PHYSICAL CONTACT

So far, we have discussed viruses we inhale, viruses we eat, and viruses we get from mom. In each case, these viruses can be spread by mechanisms over which we have little or no control. For example, unless you live like Howard Hughes, you are almost certain to have someone sneeze nearby and infect you with one of the respiratory viruses. In contrast, this lecture will focus on three viruses (HIV-1, herpes simplex virus, and human papilloma virus) that in most cases can be avoided, because they are usually spread by intimate physical contact. The fact that these viruses thrive in human populations is a demonstration of how strong the urge for intimate contact is—and how clever these viruses are to have evolved to take advantage of these urges.

HIV-1—AN URBAN VIRUS

Human immunodeficiency virus type one (HIV-1) is a close relative of HTLV-I. However, these two viruses have very different lifestyles, which result in stunning differences in the pathological conditions they can cause.

Viral Reproduction

HIV-1 and HTLV-I are both retroviruses, and the basic strategies they use to replicate their single-stranded RNA genomes are very similar. Consequently, the figures that depict HTLV-I reproduction in the last lecture can be used without much modification for HIV-1. There are, however, several very important differences.

Both viruses infect cells that have CD4 proteins on their surfaces. However, although the HIV-1 envelope protein binds to CD4, available evidence indicates that CD4 is not the HTLV-I receptor. Moreover, CD4 binding by HIV-1 is not enough for viral entry: HIV-1 must also bind to co-receptor molecules on the surface of its target cell. These co-receptors facilitate efficient fusion of the viral envelope with the cell's plasma membrane, allowing the virus to enter the cell's cytoplasm. The co-receptors used most frequently by HIV-1 are proteins that normally function as receptors for a class of cytokines known as chemokines (e.g, the chemokine receptors, CCR5 and CXCR4). Although entry of HTLV-I into infected cells has not been studied nearly as carefully as HIV-1 entry, no co-receptor for HTLV-I has been identified.

Another important difference between HIV-1 and HTLV-I is that the HIV-1 genome encodes at least four "accessory" proteins which have no HTLV-I counterparts.

These added bells and whistles make HIV-1 the "Lexus" of retroviruses. One of the HIV-1 accessory proteins, Vif, can increase the infectivity of HIV-1 about 1,000-fold. HTLV-I's lack of a Vif protein may contribute to the relatively low infectivity of that virus. HIV-1's accessory proteins (e.g., Vpr) also help make it possible for this virus to integrate its proviral DNA into the chromosomes of some cells (e.g., macrophages) that are not proliferating. In contrast, HTLV-I is limited to infecting proliferating cells. It is important to note, however, that HIV-1 cannot integrate its proviral DNA into the chromosomes of non-proliferating CD4$^+$ T cells.

Although HIV-1 can integrate its proviral DNA into the chromosomes of some resting cells, few or no new virus particles are produced unless HIV-1-infected cells are proliferating. The reason is that for new viruses to be produced, integrated proviral genes must be transcribed into RNA by the cellular RNA polymerase. And when cells are resting, the goodies required for transcription are only produced at "maintenance" levels. What this means is that the proviral genome is "silent" in most infected cells—because the majority of CD4$^+$ cells are not proliferating. However, if infected cells are activated and begin to proliferate, transcription of viral genes can increase 1,000-fold, and the program for producing new viruses can be initiated. Once this happens, it only takes about twelve hours before new virus particles begin to bud from the infected cell.

Another important difference in the reproductive lifestyles of HIV-1 and HTLV-I is that HIV-1 frequently kills the cells it infects, whereas HTLV-I rarely does. How HIV-1 kills these cells is not known. It may be the direct result of an unidentified activity of one of the accessory proteins, or it may be that the huge number of viruses budding through the cell membrane simply shred the cell as they exit.

How HIV-1 Spreads

The current thinking is that HIV-1 probably "jumped" from infected chimpanzees to humans in Africa several times around the middle of the twentieth century. This likely occurred when infected chimpanzee blood was transferred to the bloodstream of humans during the butchering of chimpanzees for food. The chimpanzee ancestor of HIV-1 is thought to be a very old virus that evolved long ago to infect nonhuman primates without causing damaging symptoms. Because humans didn't just begin butchering chimpanzees in the middle of the twentieth century, it's probable that humans had been infected with HIV-1 on many earlier occasions—but that the virus just didn't "catch on" in the human population until very recently. What has changed that made it possible for this chimpanzee virus to cause such worldwide human devastation? To answer this question, we need to examine the way that HIV-1 spreads.

As with HTLV-I, HIV-1 spreads within an infected individual by three main avenues: infection of CD4$^+$ cells by virus particles, proliferation of infected cells harboring integrated proviral DNA, and when infected and uninfected cells fuse. However, infection with HIV-1 virions is much more efficient than with HTLV-I virions, so infection of CD4$^+$ cells by virus particles is probably the major strategy used by HIV-1 to spread within its infected host. This is especially true early in infection, because cell-cell fusion (syncytia formation) usually is not observed until the late stages of an HIV-1 infection.

Because both HTLV-I and HIV-1 infect white blood cells that have the CD4 protein on their surfaces, you'd expect that HIV-1 virus and infected cells would be found predominately in the blood, semen, vaginal secretions, and breast milk of infected individuals—just as with HTLV-I. And this is true. However, in contrast to an HTLV-I infection, in which these bodily fluids contain large numbers of infected cells but very little infectious virus, the bodily fluids of an individual infected with HIV-1 contain both infected cells and infectious virus. In fact, during the asymptomatic phase of an HIV-1 infection, about one in 10,000 CD4$^+$ T cells are infected, and there are roughly 10,000 viruses circulating per ml of blood.

Based on these considerations, you would predict that the natural routes of infection would be from mother to child, as well as by sexual contact. With HTLV-I, we saw that mother to child transmission during breast feeding was a major route for viral propagation. This mode of spread was made possible by the fact that the vast majority of children infected with HTLV-I at birth experience no disease symptoms throughout their lives, making it possible for them to transmit the virus to their young and also to their mates during a lifetime of sexual activity. This is not the case for HIV-1, since children infected with HIV-1 at birth almost never survive long enough to spread the virus by sexual contact. So although about 25% of the children of HIV-1-infected mothers will be infected, mother to child transmission represents a dead end for this virus.

Because HIV-1 virions and HIV-1-infected cells are present in seminal fluid and vaginal secretions, HIV-1 can also be spread during sex. However, sexual intercourse is not an inherently efficient way to spread HIV-1.

In fact, the probability that an otherwise healthy man, who contracted HIV-1 from a contaminated blood transfusion, will infect a perfectly healthy female during a single episode of vaginal intercourse is only about one in a thousand. And the likelihood that a similarly infected woman will infect a perfectly healthy man is even smaller. What this means is that if HIV-1 were only transmitted either during childbirth or by men and women who had only one or a few sexual partners during a lifetime, HIV-1 could never become established in the human population—transmission is just too improbable. I think this explains why HIV-1 has only become a health problem in the last few decades: Lifestyles have changed recently to include what we might term "urban practices." These changes have made it possible for HIV-1 to survive, and in some settings, to thrive. What are these changes?

Epidemiologists who study the AIDS epidemic have identified three principle core groups from which HIV-1 infections now spread: female prostitutes and the men who use them, men who engage in anal sex, and intravenous drug abusers. In Africa and in parts of Asia, most AIDS cases emanate from female prostitutes and their patrons. As large population centers emerged in Africa during the twentieth century, it became much more common for men to visit prostitutes. Many of these prostitutes are infected with the "usual" sexually transmitted diseases such as herpes, syphilis, gonorrhea, and *Chlamydia*. Some of these diseases cause ulceration of the tissues of the sex organs—a condition that makes transmission of HIV-1 much more likely. Others cause inflammation of vaginal and penile tissues, resulting in the recruitment of many activated $CD4^+$ cells to these areas, greatly increasing the number of local targets available for infection by HIV-1. So although by itself, HIV-1 is not spread efficiently during sex, in the context of other sexually transmitted diseases, the likelihood of sexual transmission of HIV-1 increases dramatically. It is probable that the prevalence of the usual sexually transmitted diseases in the prostitute population set the stage for the efficient transmission of HIV-1 once prostitutes became infected with this virus. Customers of these prostitutes, equipped with both the usual sexually transmitted diseases and HIV-1, could then transmit these diseases efficiently to their wives and other sex partners. In parts of Africa, this dissemination is facilitated by the practice of "dry sex" in which the female partner uses rags or other means to dry the walls of her vagina before sex. Such un-lubricated sexual intercourse increases the chance for microtears in the mucosal barrier of the vagina, expediting viral entry.

A second core transmission group is men who practice anal sex. Apparently, Mother Nature designed the anus to have things come out of it, rather than to have things put into it, because the mucosal tissues that line the rectum are prone to tearing during anal intercourse. Disruption of the integrity of the mucosal barrier increases the probability that HIV-1 can be transmitted, especially from the insertive to the receptive partner. In addition, men who practice anal sex tend to have many, frequently anonymous, sexual partners. Not only does this promiscuity increase the probability of spreading HIV-1 more widely, but the high multiplicity of sexual partners also increases the risk of contracting other sexually transmitted diseases which, as we have discussed, can increase the chances of transmitting HIV-1. In the United States, it is estimated that among men, over 60% of HIV-1 infections were acquired during anal intercourse. Once infected, these men can spread HIV-1 infections not only to other men, but to women through vaginal or anal intercourse.

Intravenous drug abusers represent the third major focus from which HIV-1 infections emanate. Many abusers share needles or syringes, and because persons infected with HIV-1 have large numbers of viruses and infected cells in their blood, this practice also results in the efficient "sharing" of HIV-1 infections. Indeed, about half of the women in the United States who are infected with HIV-1 contracted the virus during intravenous drug abuse. People who abuse drugs also tend to be sexually promiscuous, giving the virus an additional route of infection within this core transmission group, as well as a way to radiate outwards to infect the general population.

Of course, the three core transmission groups are not independent. For example, intravenous drug use frequently is the route by which HIV-1 is introduced into the population of female prostitutes, and a relatively high proportion of men who practice anal sex also inject drugs. In the United States, roughly 90% of the men who are infected with HIV-1 contracted the virus either by intravenous drug abuse or during anal intercourse, and of these, almost 10% are members of both core transmission groups.

From these three foci of infection, HIV-1 spreads to persons who are not members of these core transmission groups, mainly by intimate contact. In addition to sexual dissemination, HIV-1 can be transmitted efficiently by contaminated blood used for transfusions or by contaminated blood products (e.g., clotting factor VIII). In fact, a person who receives a single unit of HIV-1-contaminated blood has greater than a 60% chance of becoming infected. In this country, since 1985, blood and blood products have been screened for HIV-1, dramatically reducing viral spread via this route. However, in parts of the world where blood is not carefully screened or where needles and syringes are reused, the transfer of contaminated blood is one way HIV-1 can be efficiently spread from foci of infection into the general population.

So in contrast to HTLV-I, which thrives in small, isolated tribes, HIV-1 is a virus that prospers in an urban setting. Indeed, the "modern urban practices" of prostitution, anal sex, and intravenous drug abuse have allowed HIV-1 infections to "catch on." Further, urban life provides a large number of uninfected hosts who live in close proximity with core transmission groups, making it more "convenient" for members of these groups to spread the virus to the general population.

It is important to recognize that although the three core transmission groups represent reservoirs of HIV-1 infection, these groups are very small. For example, men who practice anal sex or people who are intravenous drug abusers make up only a small fraction of the population of the United States. Further, although the flame of HIV-1 infection burns brightly in these core groups, there is only one route through which HIV-1 can spread with reasonable efficiency from these "hot spots" to the rest of the "forest": sexual promiscuity. Even in Africa—where HIV-1 can still be transmitted by contaminated blood products, needles, and syringes—over 90% of HIV-1 transmission occurs during sexual intercourse.

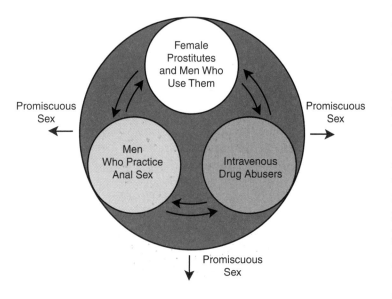

Viral Evasion Tactics

How is it that this virus is able to establish a lifelong infection in humans? The answer is that HIV-1 uses a combination of "defensive" strategies which allow it to evade the host's immune system, as well as "offensive" strategies by which it actively subverts the very host defenses that are trying to defend against it.

Once HIV-1 proviral DNA has been integrated into the chromosomes of infected cells, it can just sit there. Killer T cells do not recognize these cells as infected, so the viral genetic information is safe from attack. In this respect, HTLV-I and HIV-1 infections are very similar: Both viruses establish a reservoir of infected cells which cannot be detected by the host immune system.

At this point, however, the lifestyles of these two viruses diverge dramatically. When, in response to activation signals, infected CD4$^+$ cells begin to proliferate, the usual outcome for HTLV-I is a continued "stealth" posture, in which the integrated proviral genome is quietly passed down to the resulting daughter cells, and relatively little new virus is produced. Indeed, most HTLV-I-infected cells that begin to assemble new virus particles are rapidly destroyed by a strong immune response before viral reproduction can be completed. In contrast, when cells latently infected with HIV-1 proliferate, the more common scenario is that viral RNA transcription is stimulated, and many new virus particles are produced. Of course, the production of viral proteins "blows the cover" for these infected cells, and they now become targets for killing by T cells. However, the accessory proteins encoded by the HIV-1 genome make production of new viruses so efficient that many activated CD4$^+$ cells will crank out their load of viruses before they can be destroyed. Further, because thousands of new viruses are made per infected cell, and because these newly made viruses are so infectious (again, thanks at least in part to those accessory proteins), many new CD4$^+$ cells will be infected before neutralizing antibodies can deal with the new blast of virus.

So although HIV-1 can use latently infected cells as hidden virus reservoirs, in the body of an infected individual there is so much virus production and so many infected CD4$^+$ cells that an HIV-1 infection really should <u>not</u> be considered a latent infection. Rather it is now appreciated that HIV-1 causes a chronic infection in which huge amounts of virus are produced, large numbers of CD4$^+$ T cells are continuously killed and replaced, and in which the host immune system is engaged in a heroic battle to control and eliminate the virus. The fact that the chronic phase of an HIV-1 infection is relatively asymptomatic even though there are a large number of viruses circulating in the blood can give HIV-1 a decade-long "window" during which the virus can be transmitted to new hosts.

During each round of viral replication, HIV-1's reverse transcriptase enzyme must produce a DNA copy of the HIV-1 genome. However, this enzyme makes about one error (mutation) each time it copies a piece of viral RNA, so essentially all viruses produced in an HIV-1-infected cell are different from the virus that originally infected that cell. In fact, an untreated person who is infected with HIV-1 generally has in his body a collection of HIV-1 genomes that contains every possible single base mutation. In contrast, HTLV-I spreads within its infected hosts mainly by the proliferation of infected cells—a mode of amplification that does not require new rounds of viral replication. Because HIV-1 and HTLV-I employ different strategies for spreading within infected hosts, the effective mutation rate of HIV-1 is much greater than that of HTLV-I.

Mutations introduced into the HIV-1 genome due to errors in reverse transcription can produce three different results. First, the mutations may not change viral structure or function at all. Second, the mutations may actually kill the virus, because they disturb some essential function (e.g., the mutation might change the reverse transcriptase so that it no longer can copy viral RNA). Finally (and here's the problem), the mutations may help the virus spread more easily or evade immune detection. HIV-1 can mutate, for example, so that a viral protein that formerly was recognized by killer T cells no longer can be recognized, or no longer can be presented by the MHC molecule that the killer T cells were trained to focus on. When such mutations occur, that clone of killer T cells will be useless against cells infected with the mutant virus, and a new clone of T cells which recognizes the mutant will have to be produced. Meanwhile, the mutant virus, which has escaped surveillance by the obsolete killer T cells, is reproducing like crazy. And every time it infects a new cell, it mutates again. The bottom line is that the HIV-1 mutation rate is so high that the virus can effectively stay one step ahead of killer T cells or antibodies directed against it.

So two of the properties of HIV-1 that make it especially difficult for host defenses to deal with are its ability to establish an undetectable, "stealth" infection, and its high mutation rate. But these "defensive" strategies are only half the story.

Both HTLV-I and HIV-1 infect cells of the immune system that display the CD4 surface protein: dendritic cells, macrophages, and helper T cells. However, in contrast to HTLV-I, which usually establishes a rather benign presence in these cells, an HIV-1 infection either disrupts their function, kills the cells outright, or makes them targets for destruction by killer T cells. As a result, an HIV-1 infection damages or destroys the very cells that are required to activate the immune system and to provide killer T cells with the help they need to function. Even more insidiously, HIV-1 can turn the immune system against itself by using processes that are essential for immune function to spread and maintain the viral infection. Here's how this works.

For the adaptive immune system to be activated, an antigen (e.g., a part of a viral protein) and the T cells and B cells which recognize that antigen must meet. This is a potential problem, because only a tiny fraction of all the B and T cells circulating throughout the body will recognize a given antigen. So to increase the probability that a T or B cell will find its antigen and be activated, Mother Nature created lymph nodes—"dating bars" in which T cells, B cells, antigens, and the cells that present these antigens congregate. T and B cells circulate through lymph nodes on a regular basis, and antigens can be carried to the nodes by the lymphatic system that drains our tissues. In addition, dendritic cells stationed out in the tissues can pick up antigens and carry them into nearby lymph nodes. As a result, lymph nodes provide an environment in which there is a high concentration of immune system cells and antigens—just the right conditions to favor activation of the adaptive immune system. However, bringing all these cells and antigens together in lymph nodes can actually work to the advantage of HIV-1.

HIV-1 can attach to the surfaces of dendritic cells and be transported by these cells from the tissues, where there are relatively few CD4$^+$ cells, into the confined environment of the lymph node, where huge numbers of CD4$^+$ T cells are located. Now the virus is in business. Not only are there lots of CD4$^+$ cells within easy reach, but many of these cells are proliferating, making them ideal candidates for HIV-1 "factories." As far as HIV-1 is concerned, a lymph node is about as close to heaven as a virus can get.

Likewise, during a viral infection, virus particles that have been tagged (opsonized) by antibodies or complement proteins are collected and displayed in lymph nodes by follicular dendritic cells. This display is intended to help activate B cells which circulate through the forests of follicular dendritic cells, looking for antigens to which their receptors will bind. However, CD4$^+$ T cells also pass through these forests, and as they do, they can be infected by HIV-1 particles that are attached to the dendritic cell "trees." Because virus particles typically are displayed on follicular dendritic cells for months, lymph nodes actually become reservoirs of HIV-1. So by choosing to infect CD4$^+$ cells, HIV-1 takes advantage of the normal trafficking of immune system cells through lymph nodes, and turns these "dating bars" into its own playground.

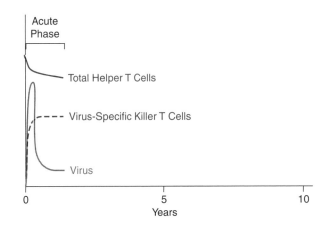

The Pathological Consequences of an HIV-1 Infection

By far the largest group of immunodeficient humans acquired their deficiency when they were infected with HIV-1. Today, nearly forty million people worldwide are infected.

An HIV-1 infection begins very much like many other viral infections. Viruses in the initial inoculum penetrate the mucosal barrier and infect CD4$^+$ cells that are found in the tissues below mucosal surfaces. The viruses reproduce in these cells, using the infected cells' biosynthetic machinery, and newly-made viruses then infect other CD4$^+$ cells to amplify the infection. Early on, the virus multiplies relatively unchecked while the innate system gives it its best shot, and the adaptive system is being mobilized. After a week or so, the adaptive system starts to kick in, and virus-specific B cells, helper T cells, and killer T cells are activated, proliferate, and begin their carefully orchestrated counterattack. The "big gun" in most first-time viral infections is the killer T cell, because each virus-infected cell may produce thousands of new viruses, and killer T cells can destroy these infected cells before viruses have a chance to multiply inside them.

This "acute" phase of the infection is frequently accompanied by flu-like symptoms (fever, chills, muscle aches, etc.) which are caused primarily by interferon and other cytokines produced by the innate and adaptive immune systems. Because lymph nodes harbor large numbers of CD4$^+$ cells, they become centers of infection. The resulting swelling of the lymph nodes (lymphadenopathy) is one of the most common early symptoms of an HIV-1 infection.

In the initial stages of the infection, there is a dramatic rise in the number of viruses in the body (viral load) as the virus multiplies in infected cells. This is followed by a marked decrease in viral load as virus-specific killer T cells and antibodies go to work.

With some viruses, the acute phase of infection ends with "sterilization": All the invading viruses are destroyed, and memory B and T cells are produced which can protect against a later infection by the same virus. There is some evidence that for a few, very lucky individuals, an HIV-1 infection may also end in sterilization. However, for the vast majority, infection with HIV-1 leads next to a "chronic" phase that can last for ten or more years. During this time, a fierce struggle goes on between the immune system and the virus—a struggle that, unfortunately, the virus always seems to win.

During the chronic phase of infection, viral loads are smaller than at the height of the acute infection, but the blood of a chronically infected person usually still contains between 1,000 and 100,000 virus particles per ml. The number of virus-specific helper T cells and killer T cells also remains high—a sign that the immune system is still trying hard to defeat the virus.

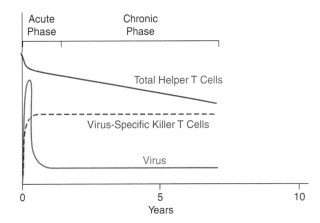

So during an HIV-1 infection, there is a cycle of birth and destruction. Each day, over a billion new viruses are produced, most of which are subsequently "cleared" by antibodies that tag the viruses for destruction by phagocytic cells (e.g., macrophages). Many new CD4$^+$ T cells are also produced each day, only to be infected by newly made viruses. Activated by the battle, these CD4$^+$ T cells produce more viruses and are destroyed either by the immune response or by the viral infection itself. To replace these cells, new CD4$^+$ cells are produced, and the cycle continues.

As the chronic phase wears on, the total number of helper T cells slowly decreases. In addition, antigen presenting cells (e.g., dendritic cells), which are required for activating both helper and killer T cells, also are infected by HIV-1, resulting in the death or malfunction of these important cells.

Eventually there are not enough helper T cells remaining to provide the help needed by virus-specific killer T cells. When this happens, the viral load increases dramatically, because there are too few killer T cells left to cope with newly infected cells.

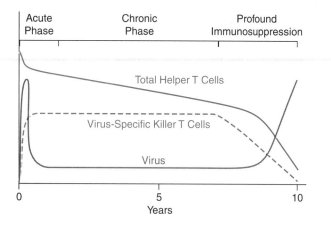

This precipitous increase in viral load usually occurs eight to ten years post infection, although in about 5% of the cases, it happens within the first two years. In the end, the immune defenses are overwhelmed, and the resulting profound state of immunosuppression leaves the patient open to unchecked infections by pathogens that would normally not be the slightest problem for a person with an intact immune system. Sadly, these "opportunistic" infections can be lethal to an AIDS patient whose immune system has been destroyed. Indeed, more than a

hundred different opportunistic infections with viruses, bacteria, parasites, and fungi have been associated with AIDS. One little-recognized consequence of the immunocompromised state of AIDS patients is that these individuals can also act as reservoirs for pathogens such as tuberculosis, which can then be spread to persons who are not at risk for HIV-1 infection.

The systematic destruction of the immune system of a person infected with HIV-1 is in striking contrast to the situation with HTLV-I. Because HTLV-I does not attack and destroy cells of the immune system, the immune systems of those who are infected with this virus remain largely intact.

In addition to being open to almost every infection you can imagine, patients in the latter stages of an HIV-1 infection also can experience a variety of neurological problems. Although the basis of these syndromes is not well understood, it is believed that they result either directly from the infection of brain cells by the virus, or from the action of various cytokines (e.g., tumor necrosis factor) which are produced by the immune system in response to infected brain cells.

Finally, persons infected with HIV-1 are more susceptible to several forms of cancer which otherwise are rarely seen. For example, B cell lymphomas are about 100 times more common in AIDS patients than in the general population. Also, in the United States, about 20% of AIDS patients contract Kaposi's sarcoma. It is believed that increased susceptibility to these cancers is not due to HIV-1 *per se*, but to the immunosuppression and the chronic stimulation of immune system cells that result from the HIV-1 infection. A herpes virus (called Kaposi's sarcoma-associated herpes virus or human herpes virus eight) has been found in over 95% of the Kaposi's sarcomas examined, so it is presumed that infection with HHV-8 plays some role in this cancer. Probably the HIV-1-associated immunosuppression either allows infection by this virus or permits reactivation of a preexisting HHV-8 infection.

HERPES SIMPLEX—A VIRUS THAT HIDES

Although HIV-1's ability to establish a latent infection gives it a reservoir from which new viruses can emerge to sustain the infection, the virus' main strategy is to attack the immune system head on. In fact, it is now clear that HIV-1 spends relatively little time "hiding" and most of the time attacking. In contrast, herpes simplex virus, after the initial infection, spends the majority of its life in hiding and none of its time actively attacking host defenses. Yet by maintaining a low profile, and by taking advantage of humans' proclivity for intimate physical contact, herpes simplex virus has had great success. This virus currently infects about a third of the human population.

Viral Reproduction

There is nothing simple about herpes simplex virus. This large, double-stranded DNA virus has enough genetic information to code for over eighty proteins. In comparison, several other DNA viruses get the job done with only about a half dozen genes. Many of the "extra" herpes genes are responsible for three of the most important characteristics of a herpes simplex infection: the ability to reproduce in cells that are not growing, to rapidly produce a burst of new virus in infected epithelial cells, and to establish a latent infection in cells of the central nervous system.

The DNA genome of herpes simplex virus is enclosed in a protein capsid. In addition, this viral capsid is coated with an amorphous layer of proteins called the tegument—a structure that is unique to herpes viruses. This tegument isn't just padding for the viral capsid, because a number of the tegument proteins perform functions which are critical for the takeover of the infected cell. Finally, the tegument-coated capsid is enclosed in an envelope supplied by the infected cell.

To initiate an infection, proteins on the viral envelope bind to heparan sulfate, a carbohydrate that is attached to proteins on the outer surfaces of every cell. This binding brings the virus close to its target so that other proteins on the virus envelope can bind to co-receptor proteins on the cell surface. Next, through the concerted action of several different viral proteins, the viral envelope fuses with the cell membrane. During the fusion process, the viral envelope is lost, and the partially uncoated virus particle enters the cytoplasm of the cell. The virus particle then makes its way to the cell nucleus where replication of the viral genome will take place. When it reaches the nuclear membrane, the viral capsid "opens," and the viral DNA enters the nucleus. At the same time, some of the viral tegument proteins also enter the nucleus where they work their magic.

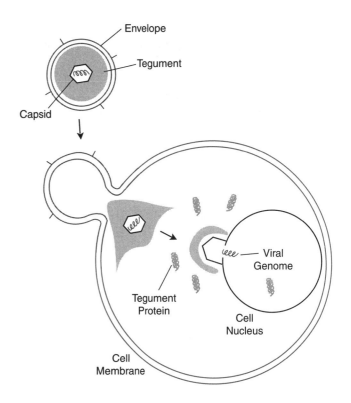

Inside the nucleus, the linear viral DNA is circularized, early viral mRNAs are transcribed, and the DNA is copied by a "rolling circle" strategy that produces many copies of the viral genome joined together on a long, linear DNA molecule. This giant piece of DNA is used as a template for further viral mRNA synthesis before being cleaved to produce many copies of the linear viral genome. These genomes are encapsidated within the nucleus, and then, after tegument proteins are added, they bud through the nuclear membrane. Currently it is not clear what happens next. Either the virus particle is transported more or less directly to the cell surface; or the envelope acquired during budding through the nuclear membrane is removed, and another envelope is picked up by budding through another membrane within the cell. In any case, completed virus particles, each composed of viral DNA, a protein capsid, tegument proteins, and an envelope are finally released from the cell.

Many of the supplies required for DNA replication are not available or are present in limited amounts in human cells that are not proliferating. Since this includes most of the cells in a human, DNA viruses have evolved two different strategies to make it possible for them to reproduce in "resting" human cells.

Some DNA viruses (e.g., adenovirus) encode proteins that force resting cells to proliferate, thereby insuring that the enzymes and building blocks required for efficient viral reproduction will be available for their use. Herpes simplex virus employs a different strategy. The genome of this virus encodes enzymes required for DNA replication which are limiting in resting cells. For example, in addition to encoding its own DNA polymerase, herpes simplex virus makes its own version of an enzyme, thymidine kinase, which it uses to help produce large quantities of the building blocks required for constructing viral DNA. Indeed, when herpes simplex virus infects an epithelial cell, it turns off most host protein synthesis, and stages a complete takeover of the cell's biosynthetic machinery. Within this "hijacked" cell, viral reproduction is rapid, and within twenty-four hours, a single infected cell can produce thousands of new viruses. This massive amount of viral reproduction isn't so great for the herpes-infected cell, and it eventually dies from the exertion.

In addition to a "productive" infection in which herpes simplex virus turns epithelial cells into doomed virus factories, this virus can also establish a stealth, latent infection in cells of the central nervous system. This latent infection usually lasts a lifetime, and from time to time these latently infected cells can "reactivate" to produce more virus.

Strategy for Viral Spread

Two closely related herpes simplex viruses exist, HSV-1 and HSV-2. The genomes of these two viruses are about 50% identical, and it is believed that they diverged from a common ancestor about 9 million years ago. By the time they are old enough to vote, about 60% of all Americans will have been infected with HSV-1, whose preferred targets are the skin around the lips and the mouth. HSV-2 predominantly infects the genital area, and about 25% of the American population is infected with HSV-2. Although each virus has its favorite sites of infection, with the breakdown of taboos against oral sex, things have gotten rather mixed up. Now roughly 40% of all new cases of oral herpes are caused by HSV-2, while about the same fraction of new genital herpes cases involve HSV-1.

Herpes simplex virus spreads when virus produced in the epithelial cells of an infected individual come in contact with the epithelial cells of a recipient. This can occur either by "skin-to-skin" contact, or when saliva or genital secretions containing the virus contact epithelial cells in abraded skin (skin in which the protective layers of keratinized cells are broken to allow access to the underlying epithelial cells) or in genital or oral mucosa. This transfer is quite efficient: It is estimated that a woman having sex with a man with active herpes has an 80% chance of contracting the disease during a single encounter.

Although oral and genital herpes infections are the most common, any region of the skin or mucous membranes can be infected. This is because the receptors to which HSV-1 and HSV-2 bind are found on most of the epithelial cells that underlie the skin or which line mucosal surfaces. Presumably, if humans enjoyed rubbing elbows as much as they like kissing and having sex, "elbow herpes" would be a common disease. Indeed, doctors who examine herpes lesions without the protection of latex gloves can contract herpes infections of the fingertips. Because there are so many epithelial cells in the skin and mucosa, the virus has many targets for infection, supporting dictum that "viruses like large organs."

This brings up another important feature of herpes simplex infections: This virus usually stays where you put it. For example, if you touch a herpes lesion with an ungloved hand, you may contract a herpes infection—but if you do, it will be on the very finger that made contact with the infected lesion. Likewise, herpes simplex virus doesn't infect just any old nerve cell. It infects nerve cells which are in close proximity to infected epithelial cells. So in contrast to HIV-1 infections, which spread to CD4$^+$ throughout the body, herpes simplex infections are usually very localized. This provincialism is certainly in the virus' best interest, because a systemic herpes simplex infection would almost certainly kill its host before the virus could be spread to another human.

The reason why herpes simplex usually doesn't produce a systemic infection is very interesting. The cellular receptor to which the virus binds, heparin sulfate, is found on the surface of almost every cell in the body. So in this respect, it might seem that herpes simplex virus is well equipped to "go systemic." However, heparin sulfate is also a major constituent of the extracellular matrix, the structural framework within which our cells reside. Consequently, because there is so much heparin sulfate surrounding our cells, herpes simplex viruses usually can't travel very far before being "soaked up" by binding to extracellular matrix heparin sulfate molecules. Only in cases when the virus, unchecked by host defenses, reproduces to overwhelming numbers does a systemic infection result.

Evasion of Host Defenses

The herpes simplex genome includes a large number of genes, roughly 50% of which are involved in evading host defenses. For example, one way the innate system deals with viruses is to deposit complement proteins on their surfaces. These proteins facilitate ingestion and destruction of the viruses by phagocytic cells (e.g., macrophages), and can destroy some viruses by poking holes in their protective envelopes. Herpes simplex virus, however, is prepared for this defense. One of the viral proteins that makes up the herpes envelope (gC) can bind to complement proteins deposited on the surface of the virus and disrupt their function.

Once the adaptive immune system has been activated, antibodies that recognize herpes simplex virus are produced. These antibodies should be able to tag (opsonize) the virus for phagocytic ingestion, providing "backup" for the complement proteins in this role. However, herpes simplex virus has proteins on its surface that bind to the tail region of IgG antibodies—the part of the antibody which normally would attach to a phagocytic cell. As a result, the antibody can no longer form a "bridge" between the virus and the phagocyte. Because phagocytosis of opsonized virus is one of the major ways the immune system deals with viruses that are outside of cells, interfering with opsonization by both complement proteins and IgG antibodies is a very effective way for the virus to evade this defense. Herpes simplex virus also has ways of avoiding destruction by killer T cells. For example, one viral protein (ICP47) reduces the quantity of viral antigens available for presentation by class I MHC molecules on infected cells. This makes it less likely that killer T cells will recognize these cells and destroy them.

Clever though they be, these viral evasion strategies ultimately are to no avail, and within about two weeks, all the infected epithelial cells are killed, either by the virus infection, or by the immune response to it. Of course, if this were all there were to the story, the virus would be in big trouble. After all, how successful would a virus be which could only infect humans (herpes simplex virus does not infect birds or animals in nature), was mainly spread by intimate contact, and that could only be transmitted during the first few weeks after infection? During evolution, it is likely that some wannabe viruses tried this strategy (spread by intimate contact, short infectious period), but they didn't survive to tell the tale.

Herpes simplex virus devised an elegant solution to the problem of continuous transmission by using two different cell types: epithelial cells in which the virus reproduces efficiently, and nerve cells in which the virus hides from host defenses. During the initial infection, herpes simplex reproduces rapidly in epithelial cells that are located at the point of physical contact. As these infected cells die, the viruses they produce are released into the surrounding tissues. There these freshly minted viruses infect more epithelial cells, amplifying the infection, and also infect nearby sensory nerve cells. Although herpes simplex virus doesn't reproduce efficiently in cells of the central nervous system, these infected neurons do provide a safe haven within which the virus is able to escape detection by the host defenses.

The details of how herpes simplex virus evades the immune system by establishing a latent infection in nerve cells are still sketchy, but several important pieces of the puzzle have been discovered. During a latent infection, few, if any, herpes-encoded proteins are produced by infected nerve cells. This is in striking contrast to the situation during a productive infection of epithelial cells in which more than eighty viral proteins are made. This paucity of viral protein production makes it difficult for killer T cells to recognize virus-infected neurons. Moreover, nerve cells normally produce relatively few class I MHC molecules, so the small number of viral proteins that are made must compete with the much larger number of normal cellular proteins for presentation on a limited number of MHC molecules. The end result is that latently infected neurons are rarely destroyed by killer T cells, so they provide a safe hiding place for the virus.

When HIV-1 establishes a latent infection in CD4$^+$ cells, the genetic information of the virus is integrated into the chromosomes of the infected cells. As a result, when these cells proliferate, the integrated proviral DNA is passed down to daughter cells, contributing to viral spread within the infected individual. In contrast, during latency in neurons, the herpes simplex virus genome exists as a "free floating" piece of DNA that is not associated with host chromosomes. This makes perfect sense. Once sensory nerve cells have taken their places in the human body, they no longer proliferate—so integration of herpes DNA into the genome of latently infected nerve cells wouldn't help the virus spread.

Of course if all herpes simplex virus did was infect and then hide, it would be impossible for a herpes simplex infection to be transmitted from human to human. So to make this all work out, the virus has devised ways of abandoning its latent state and reactivating from time to time. Virologists still do not understand how this reactivation is accomplished. Certainly, reactivation is controlled in part by viral genes that are expressed within latently infected nerve cells. However, episodes of reactivation usually are triggered by "external factors": psychological stress, physical trauma (e.g., friction during sexual intercourse), fever, ultraviolet light, etc. In any case, when reactivation occurs, the virus "wakes up," reproduces to a limited extent, exits the neuron near where it entered, and then infects epithelial cells in the vicinity to produce more virus.

So herpes simplex virus actually requires two different kinds of host cells for its life cycle: epithelial cells in which rapid viral reproduction occurs, and nerve cells in which the virus establishes a latent infection. This is in contrast to HIV-1 which can either productively or latently infect the same CD4$^+$ cell. As humans go about their normal routines of touching, kissing, and having sex, herpes simplex virus is efficiently transferred from an infected person to the epithelial cells of a recipient. Once this transfer has been accomplished, the virus deploys evasion strategies designed to hold off the innate and adaptive systems long enough to allow viral reproduction in epithelial cells and the subsequent infection of neurons. By establishing a latent infection in neurons from which the virus can be reactivated from

time to time, herpes simplex virus extends its infectious period over the lifetime of its human host. When reactivation occurs, viruses from infected neurons infect nearby epithelial cells to amplify their numbers, insuring that the virus can be efficiently transferred to the next contact. This strategy of infection, amplification, latency, reactivation, and amplification clearly works very well: Over 80% of all Americans born today will be infected with HSV-1 or HSV-2 or both.

Pathogenesis

Herpes simplex virus reproduces rapidly in epithelial cells of the skin or mucosal surfaces, destroying these cells, and producing a large burst of virus that can infect neighboring epithelial cells. From these considerations, it can be predicted that one pathological consequence of a herpes simplex infection would be the ulceration of infected epithelial cell layers caused when these cells either are killed by the virus itself or by the immune response to the infection.

From time to time, the herpes simplex virus hiding in latently infected nerve cells reactivates and infects more epithelial cells. Fortunately, because the immune response is so strong and immune memory of this virus is so good, the severity and duration of these secondary viral infections is usually limited. Thus, for most individuals, a herpes simplex infection results in an annoying (and painful), recurring skin disease.

It was originally thought that herpes simplex virus only was "shed" by infected individuals during reactivation events that led to the death of virus-producing epithelial cells and the resulting ulcers or blisters that characterize the disease. However, more sensitive measurements have revealed that virus shedding is more or less a continuous process which may or may not be symptomatic. So in a practical sense, a herpes simplex infection probably should be regarded more as a chronic infection than as a latent infection which only reactivates periodically. The fact that the infection can be transmitted by asymptomatic individuals substantially increases the probability that the virus will spread, because persons with symptomatic infections are likely to avoid intimate physical contact.

The fingers of an infected individual can spread the virus from lesions around the lips to the epithelial cells of the cornea of the eye. Indeed, about 300,000 new cases of ocular herpes are diagnosed each year in this country. The immune response rapidly destroys the infected corneal epithelial cells, and these cells are quickly replaced by the proliferation of uninfected cells. However, the virus can also infect the nerves that have endings near the surface of the eye, and a latent infection can be established. During subsequent reactivation episodes, the virus can infect fibroblasts that form the underlying structure of the eye. These infections can also be controlled by the immune response, but the resulting dead fibroblasts cannot be replaced, because the surrounding cells do not proliferate. The loss of these fibroblasts, brought about by the strong immune response to the infection, can eventually lead to blindness.

Because it is the powerful immune response that limits most herpes infections, you could predict that for people with compromised immune systems, a herpes simplex infection would result in more frequent reactivation episodes with more severe consequences. Indeed, newborns, because of the immaturity of their immune systems, are especially susceptible to devastating herpes infections. These infections most frequently occur during birth when the baby is bathed in virus-containing genital secretions from an infected mother. The risk of infection is greatest (about 30%) if the mother is experiencing a primary infection during delivery, less great (about 3%) during a symptomatic reactivation event, and very roughly 0.1% for herpes-infected mothers who are asymptomatic at the time of delivery. Neonatal herpes is frequently lethal. In addition to the "usual" sites of infection (mouth, eyes, and skin), the virus can spread, relatively unchecked by the immune system, to the central nervous system and to organs such as the adrenal glands, lungs, and liver.

The ability of herpes simplex virus to infect nerve cells suggests another possible pathological outcome of an HSV-1 infection: encephalitis. Usually, when the virus infects sensory nerve cells, it makes its way up the axon from the sensory end of the cell to the cell nucleus, where it remains, poised for reactivation and its subsequent descent back down the axon. However, the virus can also ascend the axon and infect other cells of the central nervous system and brain. This outcome is uncommon, but the results can be very serious, with a mortality rate of roughly 70% in untreated individuals.

HUMAN PAPILLOMA—A VERY QUIET VIRUS

The final bug in our Parade of viruses spread by intimate contact is the human papilloma virus (HPV). As defined by differences in their major capsid proteins, there are about a hundred different genotypes of human papilloma viruses. Each of these genotypes infects specific regions of the human body, and some are spread by activities as casual as shaking hands or walking on the deck of a swimming pool. Other genotypes usually are spread by intimate physical contact, and they will be the focus of this lecture. Although human papilloma viruses are generally thought of as "wart" viruses, it turns out that causing warts is not what these viruses usually want to do. In fact, from the standpoint of the virus, warts are an unintended and unwanted consequence of the way HPV reproduces, spreads, and evades host defenses.

Viral Reproduction

Human papilloma virus is a small, circular, double-stranded DNA virus with a single capsid made of protein. Despite its obvious medical importance, the details of how this virus reproduces are sketchy, because it has been very difficult to observe viral reproduction in the laboratory. In fact, much of what is thought to be true about HPV reproduction has been extrapolated from the habits of closely related DNA viruses which do reproduce well in cultured cells (e.g., the monkey virus, SV40). It is believed that after the HPV capsid binds to an unknown receptor on the cell surface, the viral genome somehow makes its way to the nucleus of the cell where replication takes place. After the DNA of the virus is replicated, each new viral genome is enclosed in a protein capsid, and the new virus particles leave the nucleus and exit the cell.

Because the viral genome has been sequenced, it is known for certain that this tiny virus has only eight genes. Six of these genes are expressed early after infection, and they encode proteins that are involved in regulating viral mRNA synthesis and in replicating the viral genome. The other viral genes encode the two proteins used to construct new viral capsids. Importantly, all viral messenger RNAs are transcribed from only one strand of the human papilloma virus genome. At first, it might seem that the virus has made a mistake here. After all, this tiny virus could easily have expanded its coding capacity by using both DNA strands to code for proteins—a trick that adenovirus and others use to advantage. However, as we shall see, avoiding the use of overlapping genes is critical for establishing HPV's quiet lifestyle.

HPV is extremely picky about the cells it infects and the conditions under which it will reproduce. The targets of an HPV infection are the "basal cells" that are located beneath all skin and mucosal surfaces. These cells are attached to the basement membrane, and include "basal stem cells" which proliferate on demand to replace epithelial cells as they are lost or damaged. Some of the progeny of these stem cells remain attached to the basement membrane and become stem cells themselves, whereas others are pushed upward toward the surface. Once disengaged from the basement membrane, the cells mature (differentiate) and stop proliferating.

Driven toward the surface by the continued basal cell proliferation, the maturing epithelial cells of the skin (the keratinocytes) dedicate themselves to the production of keratin proteins—much as maturing red blood cells become "factories" which produce hemoglobin. As the keratinocytes approach the skin surface, they flatten and eventually die, exhausted from the all-out effort of producing huge amounts of keratin proteins. When they die, their nuclei break down, and the cells become flattened bags of keratin. These dead cells, usually ten to twenty layers deep, function as interlocking "shingles" that provide protection against the outside environment. When, as the result of wear and tear, these dry shingles flake off the skin surface, they are replaced by a new set of shingles rising from below—the end result of basal cell proliferation. Usually it takes several weeks for the daughters of basal stem cells to reach the skin surface and become household dust.

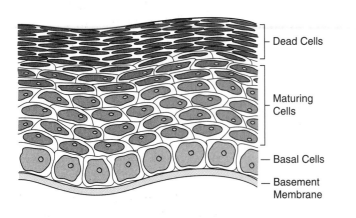

For the mucosal surface that lines the vagina, the story is a bit different. As the daughter cells are pushed higher and higher by proliferating basal cells, they flatten out, but they remain alive with their nuclei intact. In addition, the epithelial cells of mucosal surfaces do not produce huge amounts of keratin proteins, so the cells at the top of the stack (the squamous cells) are more like moist pancakes than dry shingles. Keratinized shingles make a fine covering for skin, but a dry, flaky vagina would not be so good. As the squamous cells in the upper layer slough off, they are replaced by others rising from below.

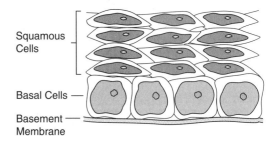

During a human papilloma virus infection, the virus makes its way through cracks in the skin or tears in the mucosal barrier until it reaches its target—the basal stem cell. Why the virus chooses these cells to infect is not completely understood, but one reason is that they are proliferating. Because HPV only has eight genes, there is no way this virus can provide all the materials required to replicate its DNA, so it must rely heavily on cells that are proliferating for replication machinery and supplies. Even so, when HPV infects basal stem cells, it produces only a small number of viral genomes per cell, and replicates more or less in sync with the replication of basal cell DNA. Moreover, the virus does not complete its reproductive program in these basal cells, and no capsid proteins are produced. So at this stage, the genetic information of the virus is "trapped" inside the proliferating basal cell.

As the infected cells move upward toward the surface and begin to mature, magic things happen. I say "magic," because virologists really aren't sure how the virus is able to change from being a "passenger" to being in control. The presumption is that as epithelial cells mature, the environment within them becomes more "permissive" for viral reproduction. In addition, two viral proteins, E6 and E7, play major roles in allowing the virus to complete its reproductive program. These proteins have the net effect of keeping the maturing epithelial cells in a "proliferative mode," so that cellular genes required for DNA replication continue to be expressed, and the "brakes," which normally would cause these cells to stop proliferating, are released. In the maturing cells, the rate of viral DNA replication increases, many more viral genomes are produced, viral capsid proteins are made, and new virus particles are assembled. By the time the virus-infected cells reach the "top floor," many new virions have been produced, and these are shed with the skin, or from the mucosal surface.

So by "shifting gears" as epithelial cells mature, the human papilloma virus goes from simply maintaining its genome in infected basal cells to a full-blown productive infection in more-differentiated epithelial cells.

When you think of it, this strategy of gently infecting basal stem cells, and waiting to produce virus until the progeny of these cells mature is really clever. Because basal stem cells are self-renewing, infecting them allows HPV to establish a long-term reservoir of infection. This strategy is similar in some ways to the herpes simplex virus' trick of using nerve cells as lifelong viral reservoirs, and infected epithelial cells as virus factories. However, in contrast to herpes, which kills the epithelial cells it infects and causes blisters, human papilloma virus does not kill its target cells, and most HPV infections are unapparent.

Viral Spread

By producing infectious virus in epithelial cells during their "elevator ride" to the surface, a continuous supply of newly made virus is delivered to the "exterior" where it can be transmitted to other humans. Human papilloma virus is spread either by direct physical contact with an infected individual or by direct contact with a surface on which the virus has been deposited. Transmission is facilitated by the fact that the viral capsid is impervious to agents that would destroy many other viruses: detergents, acids, ether, heat, and drying. In fact, the detergents used in condoms, which can damage or destroy HIV-1 and herpes simplex viruses, are helpless against HPV. Most genital HPV infections are asymptomatic and transient, so it is difficult to quantitate exactly how efficient sexual transmission actually is. However, because over a million cases of genital HPV infections are diagnosed each year in the U.S., it is clear that certain HPV genotypes (including HPV-6, -11, -16, and -18) are spread very efficiently by sexual intercourse. In the United States, the incidence of genital HPV infections has increased dramatically over the last three decades, rising at a rate faster than that of genital herpes infections. In fact, human papilloma virus is now second only to *Chlamydia* as the most commonly acquired sexually transmitted disease, with at least twenty-five million Americans currently infected. HPV types that normally infect the genitals can also be spread to the respiratory tract either during oral sex or when an HPV-infected mother gives birth.

In contrast to herpes simplex infections, which are generally lifelong, most genital herpes infections are transient, persisting for a few months to a few years. When cells proliferate, their DNA is carefully apportioned, so that each daughter cell gets a complete set of chromosomes. HPV DNA, however, is meted out to daughter cells on a random basis as infected basal cells proliferate. Consequently, a daughter cell which is des-

tined to mature and "ride the epithelial elevator" may receive all the viral genomes, while the daughter cell which is to become a new stem cell may be left "uninfected." Indeed, the likelihood of a stem cell losing the viral genome is sizable when the dividing stem cell contains only a few copies of the virus—which is typically the case in HPV infections. So one reason why human papilloma virus infections usually are not lifelong is that the viral genome can be "lost" from infected basal stem cells as they proliferate. A second reason for the transient nature of most HPV infections is that the virus is easily eradicated by the immune system. This is in striking contrast to herpes simplex virus or HIV-1 infections in which the immune system is powerless to rid the body of these viruses.

Viral Evasion Tactics

As infected epithelial cells mature and are pushed up toward the surface, they normally stop proliferating. So to keep the maturing cells in a "proliferative mode," the human papilloma virus E7 protein binds to a cellular protein, pRB, whose usual job is to put the brakes on proliferation. The binding of E7 to pRB releases this proliferation block, and leads to the activation of a whole set of genes whose expression is required for DNA replication.

But now the virus has a problem. Although expression of the viral E7 protein makes the infected cell "replication ready," the unscheduled release of the pRB-enforced replication block is something "abnormal." And when abnormal cellular DNA replication takes place, cells are programmed to commit suicide by apoptosis. Indeed, the execution of this suicide program is one of the host's best defenses against viruses which tinker with the cell's replication machinery—because when the cell commits suicide, the virus dies with it. Fortunately, the human papilloma virus has solved this problem. To keep the cells it infects from dying, HPV encodes a protein called E6, which thwarts the suicide defense. So by working together, the human papilloma virus E6 and E7 proteins can trigger unscheduled cellular DNA replication, yet avoid the usual consequence of disturbing the cell cycle: death by apoptosis.

HPV does not produce double-stranded RNA when it replicates its DNA genome; it does not wrap itself in an envelope, and all its mRNAs are transcribed from only one strand of viral DNA. Consequently, the human papilloma virus is not an interferon inducer. Because the innate system relies on interferon production as one of the earliest indications of a viral infection, the absence of interferon induction helps HPV avoid detection by the innate system. The innate system's other main indicator of a viral infection is "unusual cell death." However, HPV does not kill the cells it infects, so the innate system is not alerted in this way. In fact, because the human papilloma virus neither induces interferon production nor kills its target cells, the innate system is usually clueless that there has been an HPV infection. And if the innate system doesn't sense danger, the adaptive system doesn't get cranked up. So the current picture of a genital HPV infection is that the virus enters the skin or mucosal surfaces through cracks or tears, infects basal epithelial cells, and gently goes about its business of producing new virus, unimpeded by the adaptive or innate immune response. Then, as cracks or tears in the epithelium make further infection of basal cells possible, the virus quietly spreads within the infected individual and to his sex partners. And all this time, the immune system is "snoozing," unaware that there has been an HPV infection.

So HIV-1, herpes simplex virus, and the human papilloma virus each employs a different strategy to allow the virus to persist in the infected host long enough for efficient infection of other humans to take place. HIV-1 maintains a smoldering, chronic infection in which the virus is engaged in a continuing battle with host defenses. Herpes simplex virus chooses to "hide" in nerve cells, and to reactivate from time to time to produce infectious virus. And human papilloma virus infections are so "quiet" that the host immune system frequently "sleeps" through the early months of an infection, giving the virus time to spread while the immune system slumbers.

In most cases, the innate and adaptive systems do eventually "wake up." What alerts the host defenses is not known for certain. One possibility is that another sexually transmitted disease, for example chlamydia, causes an inflammatory reaction in the nearby epithelium, activating both an anti-chlamydia and an anti-HPV immune response. Also, some types of HPV cause raised warts, and if these are abraded, the resulting inflammation (caused, for example by superinfecting bacteria) may trigger the immune response, and blow the virus' cover. So although the human papilloma virus is most famous as a "wart" virus, causing warts is really not in the virus' best interest. Indeed, there is no evidence that HPV is spread more efficiently from the surface of a wart than from any other region of an infected epithelium.

The major player in the eventual eradication of human papilloma virus is the killer T cell. This makes sense, because antibodies can't touch the virus once it has entered its basal cell reservoir. When the immune system is finally activated, HPV and HPV-infected cells are rapidly destroyed. In fact, warts sometimes vanish in unison when the immune system finally wakes up. Importantly, this immune destruction by killer T cells is HPV genotype specific: Although genital warts may disappear, warts on other areas of the body (e.g., on the hands), caused by a different HPV genotype, usually remain. Moreover, when killer T cells awaken to destroy one genital HPV genotype, other genital genotypes may be left untouched. Because there are more than a dozen different genotypes that can infect the genital areas, people who are infected with one of these genotypes frequently are also infected, or will subsequently be infected, with other HPV genotypes. In this way, HPV can act like a "tag team" in a wrestling match: If the immune system wakes up to destroy one genotype, there often are other genotypes which are present or which will soon take its place. That's why a virus that usually establishes transient infections can end up infecting such a large fraction of the population at any given moment.

HPV-Associated Pathology

Warts are the most visible manifestation of a genital HPV infection. However, the formation of a genital wart is just the tip of the iceberg, because most genital HPV infections are unapparent. Even a microscopic histological examination frequently will not detect the infection. For example, when a large number of college-age women were tested, almost half were found to be infected with genital HPV. Nevertheless, tissues which contained abnormal cells were detected in only about 3% of those who were HPV positive, and only about 2% of the infected women had genital warts. So although human papilloma virus infections of the genitals are common, genital warts are a relatively unusual outcome of such an infection.

One of the characteristics that defines a wart is a thickening of the skin caused by an increase in the number of layers of epithelial cells between the basement membrane and the skin surface. Normally, the thickness of skin is carefully controlled. It is as if the basal stem cells "count" the number of cells above them, and divide just frequently enough to replace cells lost from the surface. Nobody knows exactly how this is accomplished, but when warts occur, it is in part because this "counting" gets confused. However, there is more to the story. Epithelial cells receive their nourishment from capillaries that are located below the basement membrane, and living epithelial cells must be located within about 0.1 mm of these capillaries to survive. This means that the living cells in a wart cannot be more than about 0.1 mm above the basement membrane—yet warts frequently are raised several millimeters above the surface of the skin. So what's a wart really made of? The answer is that a wart is composed of a thickened epithelium which has been raised by an upwelling of the basement membrane. This "bulge" in the basement membrane is caused by the proliferation of connective tissue immediately below the basement membrane and the expansion of the capillary system to support this proliferation.

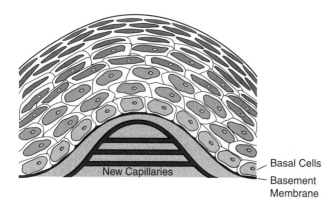

So when we ask, "How does an HPV infection cause a wart?" we are really asking, "How does the virus cause 'unnatural' proliferation of cells both above and below the basement membrane?" The answer to this question is not known for certain. Thus far, for example, none of the HPV viral proteins has been shown to cause blood vessel proliferation (angiogenesis) directly. This makes sense: If the expression of viral proteins caused abnormal cell proliferation and angiogenesis directly, all HPV infections would cause warts—but they don't.

One possible explanation for HPV-associated warts is that the cell cycle dysregulation caused by the E6 and E7 proteins increases the probability that the DNA of HPV-infected basal cells will be mutated. And these mutated cells may produce factors that encourage the proliferation of cells and blood vessels. Because only certain mutations will have this effect, only a small fraction of HPV infections should result in the production of a wart—and this is true. Moreover, warts usually appear several months after basal cells first become infected, presumably because it takes this long for the "proper" mutations to occur. So warts are likely the unintended consequence of the human papilloma virus' need to produce E6 and E7—two proteins that are essential for viral reproduction.

In rare cases, infection with certain "oncogenic" types of human papilloma virus can result in cervical cancer (mainly cervical carcinoma). I say "rare" because less than 1% of the women who are infected with genital HPV

will ever suffer from cervical cancer. However, because so many women are now infected with human papilloma virus, HPV-associated cervical carcinoma has become the second most common cancer in women worldwide.

Although over a dozen types of HPV are classified as oncogenic, it is HPV-16 which is most consistently found in cervical cancers, with HPV-18, HPV-31, HPV-45, and others being found less frequently. Clearly, human papilloma virus infections alone do not "cause" cervical cancer, since the vast majority of genital HPV infections do not lead to cancer. Yet in over 90% of the cervical cancers that have been carefully examined, one or more of the oncogenic HPV types has been detected. So although an HPV infection is not sufficient to cause cervical cancer, in most cases it appears to be a necessary element in the development of these cancers. The current hypothesis is that although an HPV infection may increase the risk of developing cervical cancer, other "insults" are required before an HPV-infected cell becomes cancerous. While the mechanism by which an HPV infection "facilitates" cervical carcinoma is not completely understood, the following is a likely scenario.

The oncogenic E6 and E7 proteins expressed in cervical cancer cells differ from their non-oncogenic counterparts in at least two respects. First, the non-oncogenic E6 and E7 proteins are relatively "weak," and usually stimulate only a small amount of "extra" cell proliferation—just enough to allow the virus to reproduce. In contrast, the oncogenic E6 and E7 proteins are more disruptive of the cell cycle than are their non-oncogenic counterparts. For example, the oncogenic E7 protein binds more strongly to the cellular pRB protein than does the non-oncogenic version of E7, and as a result, gives cells a bigger "kick" to cause them to proliferate. It's as if the oncogenic human papilloma viruses want to make absolutely sure they don't get trapped within an epithelial cell that is not proliferating. And to avoid the consequences of this bigger kick, the oncogenic E6 protein promotes the degradation of the cellular p53 tumor suppressor—a protein which normally would trigger suicide (apoptosis) in a cell whose DNA replication is really out of sync.

So the oncogenic E6 and E7 proteins are "stronger" than their non-oncogenic counterparts, but this is only half the story. In HPV-infected cells, the viral genomes usually exist as circular, double-stranded DNA molecules which float free in the nucleus of the infected cell. This is because, unlike HIV-1 or HTLV-I, the human papilloma virus does not encode an integrase protein which can facilitate the precise integration of viral DNA into cellular chromosomes. Nevertheless, on rare occasions, HPV DNA can be integrated into the cellular genome. This "unaided" integration of HPV DNA is a random event, which probably happens during the repair of damaged cellular DNA. Because integration of HPV DNA is imprecise, viral genes are usually disrupted or lost when viral DNA is pasted into a cellular chromosome. Of course, this loss of viral genetic information effectively inactivates the virus, so integration of its DNA into the genome of a cell is a dead end event for a human papilloma virus. This is in contrast to HIV-1, which makes its living by precisely integrating its proviral DNA into the genome of HIV-1-infected cells.

Although unaided integration usually results in the loss of some HPV genes, the genes encoding the E6 and E7 proteins of oncogenic HPV genotypes are almost always found intact in cervical cancer cells. Moreover, integration of these viral genes can result in the increased expression of the E6 and E7 proteins, although the reason for this isn't completely clear. In some instances, the gene for a viral protein (E2), which normally helps keep expression of E6 and E7 at relatively low levels, is deleted during the integration event. In other cases, signals within the mRNAs that code for E6 and E7, which normally would shorten the lifetimes of these mRNAs, are clipped off during integration, resulting in more-stable E6 and E7 mRNAs and higher E6 and E7 protein levels.

As long as the viral genome is not integrated into the chromosomes of an infected cell, the added strength of the oncogenic E6 and E7 proteins appears to be of little consequence. However, when the genes for oncogenic E6 and E7 are fortuitously integrated so that these proteins are expressed at unnaturally high levels, things can get ugly. Now the cell is strongly driven to mutate by the over-expressed (and more powerful) E6 and E7 proteins. And a tumor suppressor, p53, which normally would keep these mutant cells from becoming cancerous, has been eliminated by the action of E6. So it is believed that infection with an oncogenic human papilloma virus facilitates the evolution of cancer cells when viral genes, which are absolutely required for viral reproduction, are expressed "unnaturally" from an integrated viral genome. The proteins encoded by these viral genes function to increase the rate at which infected cells mutate, and to disable mechanisms that normally protect cells from these mutations.

Fortunately, integration of an oncogenic HPV genome in such a way as to deregulate E6 and E7 expression is a rare event. And that's one reason why so few women infected with HPV ever get cervical cancer. In addition, it takes more than the destruction of one tumor suppressor protein (e.g., p53) to create a human cancer cell: Additional mutations in cellular genes must occur

before a cell with integrated HPV genes can become cancerous. That's why cervical carcinomas usually arise decades after the initial HPV infection. During their "precancerous" period, HPV-infected cells may accumulate mutations and progress through stages in which they become increasingly more cancer-like. In the final stage of this progression, the mutating cells may "learn" the deadliest trick of all: how to break through the basement membrane and metastasize to other parts of the body.

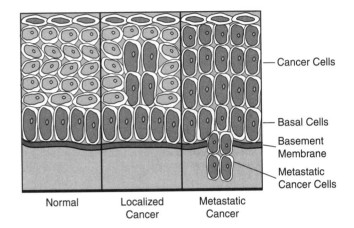

Although oncogenic human papilloma viruses can infect several other areas of the female reproductive tract (e.g., the vulva and the vagina), most HPV-associated cancers occur in the "transformation" zone of the cervix—the area where the epithelium changes from the multilayered, squamous epithelium of the vagina to the single-layered epithelium of the endocervix. Why this region is a hot spot for cervical cancer is a complete mystery. However, because cervical cancer is usually limited to this one area, and because cervical carcinoma generally develops in stages over a period of years, Pap smears taken from the transformation zone can be extremely valuable in diagnosing cervical cancer in the early stages.

It is important to note that although oncogenic HPV types can cause warts on the exterior genitals, most HPV-associated genital warts are actually caused by non-oncogenic HPV genotypes (usually HPV-6 and HPV-11). However, because it is common for individuals to be infected with more than one HPV genotype at a time, the presence of genital warts can signal a possible infection with oncogenic genotypes. Interestingly, genital herpes simplex infections were once believed to play a role in cervical cancer. However, because herpes DNA is not consistently found in cervical cancer cells, it is now thought that herpes simplex infections of the genitals, like external genital warts, are just "markers" for the types of sexual activities that can lead to infections with oncogenic HPV genotypes.

Of course, men also can be infected with oncogenic HPV genotypes (that's how women get infected, right?). However, HPV-associated penile cancer is relatively rare—perhaps because the penis has no transformation zone analogous to that of the uterine cervix. In this regard, it is interesting that a similar transformation zone does exist in the respiratory tract where the multilayered, stratified squamous epithelium of the vocal cords meets the pseudo-stratified epithelium that lines the trachea. It is this area which is the primary target for respiratory HPV infections. In adults, these infections usually involve two of the HPV genotypes (HPV-6 and HPV-11) which usually cause genital infections, and which are "transplanted"

during oral sex. Respiratory HPV infections can progress from the vocal cords downward toward the lungs, and the resulting warts can block the airways if untreated. The respiratory tract of an infant can also be infected during the birth of a child to an HPV-infected mother.

At this point, the last of our twelve model viruses has passed in review. The participants in our Parade are all established viruses that have learned to live in relative harmony with their human hosts. In the next lecture we will examine some viruses which are not so "harmonious." These are the "emerging" viruses, some of which may well be future participants in the Bug Parade.

Table 6.1 reviews the ways the final three viruses in our Parade solve their problems of reproduction, spread and evasion.

Table 6.1   Viruses We Get by Intimate Physical Contact

	HIV-1	Herpes Simplex	Human Papilloma
R E P R O D U C E	Single-stranded RNA genome	Large, double-stranded DNA genome	Small, circular, double-stranded DNA genome
	Has capsid plus envelope	Has capsid, tegument, and envelope	Has single capsid
	Retrovirus—reverse transcriptase produces double-stranded DNA copy of genome which is integrated into host chromosome	Replicates as "rolling circle"	Replicates slowly in basal epithelial cells and rapidly in more "mature" epithelial cells
	Cytolytic	Non-cytolytic in nerve cells Cytolytic in epithelial cells	Non-cytolytic
S P R E A D	Infects CD4$^+$ T cells, dendritic cells, and macrophages	Replicates efficiently in epithelial cells, but produces little or no virus in nerve cells	Infects basal epithelial cells, but produces no virus until these cells "mature" and move toward surface of epithelium
	Spreads within individual mainly when virus infects CD4$^+$ cells—causes systemic infection	Localized infection—not systemic	Localized infection—not systemic
	Spread by intimate physical contact—efficiency of spread increases dramatically in context of other sexually transmitted diseases	Spread efficiently by intimate physical contact	Spread by intimate physical contact when damage to epithelial layer exposes basal cells
	Establishes latent infection of CD4$^+$ cells	Establishes latent infection of nerve cells	Establishes chronic infection of basal epithelial cells
E V A D E	Hides in latently infected CD4$^+$ cells	Hides in latently infected nerve cells	Hides in infected basal epithelial cells
	High mutation rate	Viral proteins interfere with action of complement proteins and antibodies	"Quiet" virus—adaptive immune system usually is not alerted immediately
	Dysregulates and destroys immune system cells		Viral proteins drive cell proliferation and protect against the consequences

Lecture

7

Emerging Viruses

R E V I E W

In the last lecture we discussed three, very different viruses, all of which are spread by intimate physical contact. HIV-1 has a single-stranded RNA genome, but replicates through a DNA intermediate, whereas herpes simplex and human papilloma viruses both have double-stranded DNA genomes. Herpes simplex virus is a huge virus with so much genetic information that it can even reproduce in cells that are not replicating their DNA. In contrast, the human papilloma virus and HIV-1 are both small viruses that rely heavily on the biosynthetic machinery of the host cell to reproduce. Although HIV-1 can infect some cells that are "resting," the production of new HIV-1 particles is very inefficient unless the infected cell is proliferating. So as part of their reproductive strategies, both HIV-1 and HPV produce proteins which can force infected cells to proliferate.

All three viruses have evolved to spread by taking advantage of humans' desire for intimate physical contact. Herpes simplex virus is spread efficiently by kissing and by oral sex, but HIV-1 and the human papilloma virus are rarely spread in these ways. All three viruses can be transmitted by vaginal intercourse, but this is not an inherently efficient route for the spread of HIV-1. However, in conjunction with other sexually transmitted diseases, the efficiency of transmission of HIV-1 by either vaginal or anal intercourse increases markedly. Consequently, persons who are sexually promiscuous are not only more likely to encounter a partner who is infected with HIV-1, but are also more likely to be infected with other sexually transmitted diseases which can "facilitate" an HIV-1 infection. Transmission of HIV-1 by anal intercourse is more efficient than by vaginal intercourse, and men who practice anal sex represent a central focus for dissemination of HIV-1.

Of course, a virus that spreads by intimate physical contact, but is rapidly wiped out by the host immune response would not last long in the human population. Average humans just aren't that physically intimate. So to be successful, a sexually transmitted virus must find a way to persist in the infected individual long enough to have a high probability of being passed on to another human. HIV-1, herpes simplex virus, and the human papilloma virus all have solved this common problem, but in three different ways.

In CD4$^+$ cells, HIV-1 establishes a latent infection that is very difficult for the immune system to detect, so these cells can serve as virus reservoirs. However, from time to time, cells latently infected with HIV-1 can switch to a "productive mode," and can turn out many new virus particles. Consequently, HIV-1 is able to use CD4$^+$ cells both as virus reservoirs and as virus factories. When HIV-1-infected cells do produce new viruses, the cells usually either die from this effort, or are killed by the immune response. And because CD4$^+$ cells are among the most important cells of the immune system, the ability of the host to control the infection is slowly diminished as more and more of these cells are killed. Moreover, the HIV-1 mutation rate is so high that the virus can stay one step ahead of killer T cells and antibodies directed against it. As a result, HIV-1 actually causes a chronic infection in which huge amounts of virus are produced, large numbers of CD4$^+$ T cells are continually killed and replaced, and in which the host immune system is engaged in a heroic battle to control and eliminate the virus. During this protracted conflict, the infected individual is generally asymptomatic, giving the virus almost a decade-long "window" during which it can be spread by intimate physical contact. Eventually, the host's immune system is destroyed either by the viral infection or by the immune system's ill-fated attempt to rid the body of the virus—and the infected individual succumbs to opportunistic infections.

Herpes simplex virus has evolved a two-cell strategy to lengthen its infectious period. Initially, herpes simplex virus infects epithelial cells of the skin or mucosal surfaces, and produces many new viruses, which are immediately available for transmission to new hosts. Then, before the virus can be eradicated by the host's immune response, herpes simplex infects nearby sensory nerve cells where it "hides" by establishing a latent infection. From this dormant state the virus can be "awakened" from time to time to infect epithelial cells and produce more viruses—which can then infect new hosts during physical contact. Although these recurrent episodes of viral reproduction can result in blisters as epithelial cells are killed either by the virus or by the immune response to the infection, many reactivation events are asymptomatic. So by infecting two different cell types (nerve cells and epithelial cells), herpes simplex virus eludes the immune system, establishes a viral reservoir, and makes it possible for an infected individual to spread the virus to other humans throughout his lifetime.

Whereas HIV-1 uses the same cell type both as a reservoir and to produce new virus particles, and herpes simplex virus uses two different cell types for these purposes, human papilloma virus employs epithelial cells at two different stages of maturity to serve as reservoirs and virus factories. HPV infects basal stem cells that are attached to the basement membranes which underly all epithelial surfaces. Because these cells are "immortal," they act as long-term viral reservoirs. In these cells, viral genetic information exists as slowly replicating circles of DNA, but no new viruses are produced. Then, as the descendants of these infected basal cells mature and rise toward the epithelial surface, they become permissive for viral reproduction, and new virus particles are assembled. Finally, when these cells reach the top of the "epithelial elevator," the newly minted viruses are released and become available to infect other humans. Because human papilloma virus does not kill the cells it infects, and because it does not induce the production of interferon, the host's immune response generally "sleeps" through the first months or years of an HPV infection. During this time, new virus particles are produced continuously as infected epithelial cells mature, giving this "quiet" virus a long period during which it can spread to a new host. However, in most cases, the immune system eventually "awakens" and eradicates the HPV infection. Consequently, in contrast to HIV-1 and herpes simplex infections which are lifelong, most human papilloma virus infections are transient.

Human papilloma virus depends on proliferating cells to provide the materials required for its own reproduction. However, maturing epithelial cells normally stop proliferating after they disengage from the basement membrane. So to keep infected cells "in cycle," HPV encodes two proteins, E6 and E7, which act in concert to drive the proliferation of infected cells. Although viral reproduction is dependent upon the functions of these two proteins, the "unscheduled" cellular DNA replication they provoke can also cause the DNA of HPV-infected cells to mutate. It is likely that some of these mutations result in the production of factors which induce cell proliferation and angiogenesis, leading to the formation of a wart. In rare cases, the DNA of oncogenic HPV genotypes can be integrated into the chromosomes of infected basal cells. When this happens, the "stronger" oncogenic E6 and E7 proteins can be overproduced, and this can lead to an even higher rate of cellular DNA mutation. Sometimes, usually after many years, enough mutations accumulate to turn an infected basal cell into a metastatic cancer cell. So although human papilloma virus infections are common, they are usually unapparent. Only occasionally do HPV-induced warts appear, and only rarely do HPV-infected cells become cancerous.

EMERGING VIRUSES

There are a number of different definitions of what constitutes an emerging virus. My favorite is a simple one: An emerging virus is one that has only recently become apparent. Of course, viruses have always been emerging, but the ones we are really concerned about are the ones that are emerging now.

Where Do Emerging Viruses Come From?

Emerging viruses come from Mars. No, just kidding! Because any new virus must learn how to solve the problems of reproduction, spread, and evasion of host defenses, it is unlikely that we will see a "brand new" virus evolve from scratch during our lifetime—the life span of a human is just too short on the evolutionary time scale for such an unlikely event to take place. No, it is much more likely that any viruses that emerge during our lifetimes either will be preexisting viruses that are revealed because of changes in the "environment" in which the virus lives, or will be recent variants of viruses which have existed on earth for thousands of years.

Some viruses "emerge" when technology becomes sufficiently advanced to allow their presence to be detected. An excellent example of this is hepatitis C virus. In the late 1960s and early 1970s, hepatitis B virus had been identified as an agent which could cause hepatitis in recipients of blood transfusions. This identification was facilitated by the fact that the blood of infected individuals is usually chock full of decoy as well as infectious hepatitis B virus particles. However, it soon became clear that all cases of transfusion-related hepatitis could not be attributed to hepatitis B, and the search began for other viruses which might also be contaminating the blood supply. Nevertheless, it was not until the late 1980s that advances in biotechnology made it possible to identify hepatitis C virus and to produce diagnostic reagents which could be used to screen for hepatitis C-contaminated blood. By that time, hepatitis C virus had already infected millions worldwide.

Viruses can emerge when humans change their lifestyles. Until about 6,000 years ago, humans lived in relatively isolated "tribes." At about that time, cities began to grow up, bringing larger numbers of people into close proximity. This change in lifestyle made it possible for viruses like measles—which have short infectious periods, whose infections confer lifelong immunity, and which only infect humans—to be passed in an unbroken chain from one person to the next. Where measles "emerged from" is a mystery. Perhaps this virus originally infected some animal host. In any case, the increased population density in cities made it possible for measles to become a successful human pathogen.

HIV-1 is another excellent example of a virus which emerged due to changes in human lifestyles. Recent analysis suggests that the ancestor of HIV-1 was acquired by humans from chimpanzees around the middle of the twentieth century. HIV-1 does not cause AIDS-like symptoms in chimps, illustrating the important concept that the same virus can cause very different diseases even in species as closely related as chimpanzees and humans. Because HIV-1 has evolved a stable virus-host relationship with chimps, it is likely that HIV-1, or a virus very much like HIV-1, has existed in the chimpanzee population for a relatively long time. If this is true, HIV-1 probably "jumped" from chimps to man much earlier than the twentieth century, but just never caught on in the human population until recently. This view is supported by the fact that HIV-1 is a virus with an "urban lifestyle," and most of Africa only became urbanized in the last fifty years. As more and more Africans moved from the country into larger cities, men left their families and traveled to the cities for employment, increasing the "market" for prostitutes in these urban centers. Indeed, many African men have "second wives" in the city—wives whom they share with other workers. These second wives can act as foci from which HIV-1 infections can spread to workers and then to their families. In addition, urban growth increased the need for long-haul truck transport of various commodities across Africa—and many of the men who drive these trucks visit prostitutes along their routes, facilitating the spread of HIV-1 over great distances. So the urbanization of Africa made it possible for a chimpanzee virus to emerge as a successful human pathogen.

Viruses can emerge when humans "turn over rocks." For example, the clearing of rainforests has brought humans into contact with animal species that were rarely encountered before. The result is exposure to viruses which can infect humans, but for which a human is not the normal host. The most notorious of these are the hypervirulent viruses for which humans represent a "dead end" infection. These viruses are so lethal that infected humans frequently die before they can pass on the virus. Ebola virus first emerged from an unknown, probably animal host in the African rainforests in 1976. Marburg virus, a close relative of Ebola, was first reported in 1967 when lab workers in Marburg and Frankfurt, Germany contracted the virus while processing kidneys from infected African green monkeys obtained from Uganda. Both Ebola and Marburg can be passed from human to human by physical contact and perhaps by inhalation, and the majority of infected individuals hemorrhage and die with blood oozing from their mucosal surfaces (e.g., gums and nose). Fortunately, quarantine is usually effective in limiting the transmission of these viruses, and outbreaks are sporadic, occurring only every few years—usually when humans come in close contact with infected, nonhuman primates. Because these viruses only infect humans sporadically, other natural hosts for Ebola and Marburg must exist—hosts in which infections with these viruses probably cause few or no pathological consequences. To date, however, the natural hosts for Ebola and Marburg have not been identified.

In 1993 in the Four Corners area of northwestern New Mexico, there were a small number of unexplained deaths due to suffocation when fluid accumulated in the lungs of afflicted individuals. The cause of these deaths was quickly determined to be infection with a hanta virus that is now called the Sin Nombre (no name) virus. Since that time, several hundred cases of infection with this hanta virus have been documented in many different parts of the United States. Most of these cases occurred in rural areas, and a search revealed that the natural host for this virus is the deer mouse which is widespread in the United States and Canada. These mice can be infected with Sin Nombre virus without obvious pathological consequences, and the virus is most likely spread when mice (or humans) come in contact with virus that has been shed in the urine of infected mice. So the Sin Nombre hanta virus is a good example of a virus that emerged because humans intruded on an established virus-host system. Because the hanta virus quickly kills about 40% of the humans it infects, humans represent a dead end host for Sin Nombre infection.

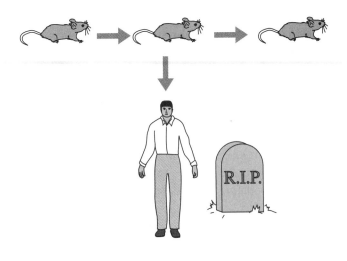

Infections of humans with viruses like the hanta virus, which have birds or animals as their natural hosts, are called zoonoses—and most emerging viral infections are zoonoses. Whereas the twelve viruses in our Bug Parade have evolved to live in "harmony" with humans, zoonotic viruses have learned to solve the problems of reproduction, spread, and evasion in their bird or animal hosts. So for zoonotic viruses, humans are irrelevant for their survival. This means, in principle, that zoonotic viruses can kill as many humans as they wish without endangering their ability to be maintained in the bird or animal population.

In the summer of 1999, a virus now known to be the West Nile virus killed seven people in the New York City area. This virus is fairly common in parts of Africa, Europe, and Asia, but only recently "emerged" in the United States. Birds are the natural hosts for the West Nile virus, and the "vector" which transmits the virus from bird to bird is the mosquito. Humans can also be infected with the West Nile virus when they are bitten by infected mosquitos, but human infection is probably a dead end for this virus: The amount of West Nile virus in the blood of an infected human is usually too small to be efficiently transmitted by a mosquito to another host.

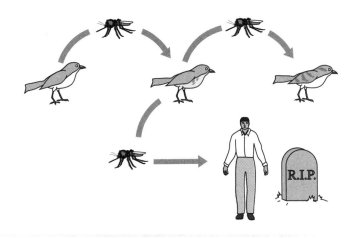

The "rock" that was turned over to allow the emergence of the West Nile virus in the U.S. was likely the transport of infected mosquitos or birds to this country from another part of the world, probably the Middle East. Indeed, world travel is responsible for the emergence of many viruses in new locations. Over a million passengers travel internationally each year, making spread by human carriers trivial. And because air travel is so rapid, infected individuals can reach their destinations before any symptoms appear. The West Nile virus can infect more than 60 bird species, including crows, blue jays, and sparrows. Having birds as hosts is a great idea for a virus, because birds, like humans, are big travelers, and can spread the virus over large geographic areas. Moreover, pools of water in discarded tires make ideal breeding grounds for virus-infected mosquitos, and these tires are routinely shipped across the country for recycling. West Nile virus has already been detected in eleven states along the East Coast, and it is predicted that it will spread as far as the West Coast during the next decade. Fortunately, by keeping the mosquito population under control, the number of West Nile infections can be kept to a minimum. Indeed, increased spraying for mosquitos in the New York area is credited with helping trim the number of deaths attributed to the West Nile virus from seven, in the summer of 1999, to only one in the summer of 2000.

Viruses can emerge when existing viruses mutate. Successful viruses coevolve with their hosts and settle down into a lifestyle that allows them to spread efficiently within the host population. However, the evolution of viruses is driven by mutation, and viral mutation does not cease even when viruses become successful pathogens. Many viruses use error-prone polymerases to replicate their genetic information. In extreme cases (e.g., hepatitis C virus and HIV-1) this results in a mutation rate so high that essentially every new virus produced will be a mutant. Viruses such as influenza, which have segmented genomes, can also mutate by exchanging gene segments with closely-related viruses that infect other species. In addition, when retroviruses like HIV-1 and HTLV-I integrate their genetic information into the genome of a host cell, it is relatively easy for these viruses to "steal" cellular genes or to recombine with other retroviruses which may have infected the same cell. The result of such events can be a mutated retrovirus with altered properties.

The ability of viruses to rapidly change their genomes gives them flexibility to adjust as their hosts "upgrade" their defenses. However, when existing viruses mutate, the result can also be the emergence of "new" viruses. For example, viral mutations can change the host range of a virus, allowing the virus to "jump" to a new species. In the early 1970s, a virus emerged which began to infect dogs in Europe. Molecular analysis showed that this virus, canine parvovirus, differed by only two small mutations from a much older virus, feline panleukopenia virus. So only two mutations were required to switch a cat virus to a dog virus. Closer to home, we have already discussed how influenza virus can jump from birds to humans when the viral genome mutates by re-assortment of gene segments during infection of a pig—an animal which hosts both bird and human flu viruses.

Mutations can also change a benign virus into a deadly one. In 1983, due to a single amino acid change, a strain of influenza virus emerged that nearly wiped out the chicken industry in Pennsylvania. This virulent strain caused a systemic and fatal infection, destroying cells in the chicken's kidneys, lungs, brain, liver, and spleen.

Can Emerging Viruses Be Dangerous?

You bet they can! So far, public attention has focused mainly on emerging viruses like Ebola or hanta that seem to strike at random and which kill a large fraction of the humans they infect. However, to my mind, these hypervirulent viruses are not the most dangerous ones. First, these viruses are not easily passed from human to human, and certainly not by casual contact. Ebola virus, for example, can be spread by infected humans, but when this happens, the infected individual is usually bleeding from the nose and gums—hardly someone who would be contacted "casually." As a result, quarantine measures are quite effective in containing outbreaks of viruses like Ebola and Marburg. In addition, because hypervirulent viruses kill humans so efficiently, their natural hosts must be non-human—hosts in which they have evolved to do little damage and to spread easily. This means that so long as humans keep their distance from these diseased, nonhuman hosts, it is unlikely that we will be infected by these zoonotic viruses. For example, the hanta virus' natural host is the mouse, and the virus spreads when mice snuffle surfaces on which infected mice have urinated. Since humans are unlikely to be snuffling the urine of other humans infected with the hanta virus, keeping the mouse population under control should prevent most hanta virus infections.

No, my idea of a really dangerous virus is not one of these hypervirulent viruses that we read so much about. I'm guessing that the killer virus we all should fear (let's call it the Andromeda Strain) will have the following characteristics. First, the Andromeda Strain will have humans as its natural host (or at least one of its natural hosts). A virus that relies on an animal or bird as its natural host, and for which humans represent only a dead end infection, would be too easy to deal with, as we have already discussed.

Second, to allow for the infection of a large fraction of the human population, the Andromeda Strain will be easily spread by casual contact. Such a virus, for example, might be spread by the respiratory route. That way, you wouldn't even have to touch the infected person to become infected—you'd just have to be nearby when he coughed or sneezed.

Third, although the virus will kill a high proportion of the people it infects, the Andromeda Strain will kill its hosts only after a long, relatively asymptomatic, chronic infection. This feature would give the virus a long window of infection, during which the virus could be transmitted.

Finally, the Andromeda Strain will probably evolve from a present-day virus which mutates rapidly. I think it's unlikely that the Andromeda Strain already exists somewhere on earth. With rapid worldwide transportation of goods and people, the probability is great that at least some humans have turned over most of the rocks, and have had a chance to "sample" most of the viruses found under those rocks. This means that we probably must wait for the Andromeda Strain to evolve. And by using an existing virus as its starting point, the Andromeda Strain wouldn't have to begin its evolution from scratch—a process which could take a very long time indeed. Further, because the evolutionary distance between a relatively benign virus and a killer virus may be great, I'm betting that the present-day virus from which the Andromeda Strain evolves will be one which mutates rapidly.

We have already seen at least one abortive attempt at this type of evolution during the last century. As we discussed in Lecture 4, influenza A is a human virus which is efficiently transmitted from human to human via the respiratory route. In addition, influenza A virus can be transmitted from fowl to fowl and from pig to pig, and sometimes these fowl or pig viruses can infect humans.

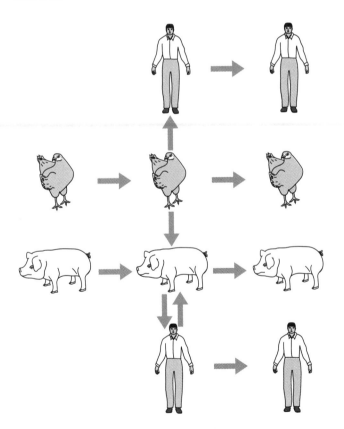

Influenza A virus employs two mechanisms to mutate rapidly. The first, which relies on an error-prone RNA polymerase, results in antigenic drift in which one or a few letters of the viral genetic information is changed almost every time the virus replicates. The second muta-

tional mechanism, antigenic shift, depends on the fact that the influenza genome is comprised of multiple RNA segments. Antigenic shift can generate dramatic changes in the viral genetic makeup when segments of the human virus are exchanged for the corresponding bird or pig segments. In 1918, mutational events occurred that changed influenza from a cold virus to a killer. Virologists still are not sure which mutations were involved, but the result was an influenza virus that killed about twenty million people. This virus was so virulent that there are reports of healthy soldiers who collapsed, and then died only one day later. Parts of the 1918 viral genome have recently been recovered from lung tissue preserved from individuals who died from the infection. Sequence analysis of these genome fragments suggests that the virulent 1918 influenza strain had both pig and bird ancestors. One current hypothesis is that an antigenic shift occurred early in 1918, producing a new strain with low pathogenicity. Then, as this strain circulated in the human or animal population, antigenic drift produced a killer virus which, instead of being confined to the airways, was capable of establishing a deadly, systemic infection.

Although the 1918 flu virus killed millions, it was not the Andromeda Strain. The world's population in those days was about two billion, so only about 1% of the population died during the influenza pandemic. This was partly because the less-virulent strain which circulated earlier in the year probably immunized many against the killer strain that came later. But more importantly, the 1918 strain of influenza A was so virulent that most who were infected died before they could distribute the virus widely to others. Consequently, although the 1918 flu had some of the properties we would predict for the Andromeda Strain—evolution from an existing human virus which has a high mutation rate and which is spread by casual contact—still, it lacked one very important attribute: a long, relatively asymptomatic contagious period. It was probably for this reason that the 1918 influenza strain "burned itself out" after only two years.

So the 1918 flu was not the Andromeda Strain—but it came very close. Imagine what might happen, for example, if influenza virus mutated to be as deadly as the 1918 strain, but killed its victims more slowly, say over a period of a few months. This would give infected individuals an extended opportunity to spread the virus by coughing and sneezing. Of course, to pull this off, the virus would have to learn to evade the immune system during this long infectious period—and that's not a trivial problem for a virus to solve. However, given the large reservoir of avian influenza virus sequences, and influenza's propensity for antigenic shift and drift, we must be vigilant lest a new variant of influenza evolves unnoticed—a virus which might indeed have the properties of the Andromeda Strain.

Our "turning over of rocks" during the urbanization of the twentieth century has produced another near miss for the Andromeda Strain: HIV-1. As we discussed in the last lecture, HIV-1 has most of the qualities of the Andromeda Strain. HIV-1 is a deadly virus that eventually kills essentially all humans it infects. It has a long, mostly asymptomatic infectious period made possible in part by its high mutation frequency, and partly because it uses the host's immune system to its own advantage and then destroys it. Currently, about .5% of the world population is infected with HIV-1, and the virus continues to spread. However, HIV-1 is not the Andromeda Strain. The reason, of course, is that it is not spread by casual contact: Usually, you have to "do something" to contract this virus. For the most part, you can avoid being infected with HIV-1 simply by avoiding intimate physical contact with someone who is carrying the virus.

Imagine, however, what could happen if HIV-1 somehow mutated to be spread by the respiratory or fecal-oral route. Then HIV-1 <u>would</u> be the Andromeda Strain! In fact, this may be the greatest danger that HIV-1 poses: the possibility that this virus may mutate to be spread by casual contact. After all, we already have discussed several examples of viruses which can mutate to change their mode of transmission. Human strains of influenza virus are spread by the respiratory route, but antigenic drift can produce a strain which, at least in ducks, is spread by the fecal-oral route. Likewise, some strains of human adenovirus are spread by the respiratory route, whereas other, closely related strains are spread by the fecal-oral route. Thankfully, changing the mode of transmission from intimate physical contact to the respiratory route is probably much more difficult than switching from respiratory to fecal-oral. However, HIV-1 mutates very rapidly, huge numbers of new (mutated) viruses are produced each day in every infected individual, and millions of humans are infected. So the emergence of strains of HIV-1 which can be spread by casual contact is not something that any virologist would rule out. And if this were to happen, humans would be in deep, deep trouble.

What Can Be Done About Emerging Viruses?

As long as existing viruses are free to mutate, and humans continue to rub shoulders with birds and animals, new viruses will emerge. Many of these emerging viruses will be zoonoses. For example, there are about two dozen nonhuman primate species which are currently infected with retroviruses related to HIV-1. Virologists must be on guard, because one or more of these viruses might emerge as a human pathogen. If, like HIV-1, these emerging viruses were to go undetected for decades, they could contaminate the blood supply, and stealthily infect a significant fraction of the human population.

A vast pool of influenza A virus genes exists in birds and other animals (e.g., pigs), and influenza can easily tap into this gene pool to create new viral strains. Virologists must detect new flu strains that are highly pathogenic, and prepare vaccines to protect humans against these emerging viruses.

The current AIDS epidemic is especially worrisome, because this virus already has so many properties of the Andromeda Strain, and because it can mutate rapidly to acquire additional, deadly characteristics. We must do whatever is possible to control the spread of this virus. Unfortunately, the most straightforward controls involve modifying human behavior—something which is notoriously difficult to do. Drugs are available which can extend the life of HIV-1-infected individuals. The mechanisms by which these drugs act have been elucidated by the same basic research that has helped virologists fathom the mind of the AIDS virus. In fact, much of this research was done on <u>animal</u> retroviruses decades before the AIDS epidemic was recognized. This illustrates how important it is to discover how viruses which <u>now</u> infect animals and birds solve the three problems of reproduction, spread, and evasion. By studying contemporary animal viruses, virologists will be much better prepared to deal with pathogenic human viruses which may emerge from them.

Although life-extending drugs are available for AIDS patients, there is no cure for an HIV-1 infection. Many virologists believe that the best way to control the spread of HIV-1 is to produce a vaccine which will protect against this virus. One great advantage of a vaccination is that, for the vaccine to be effective, the recipient does not have to make any changes in hygiene or lifestyle: He simply has to show up to be vaccinated. However, as we will discuss in the next lecture, developing an AIDS vaccine will not be easy.

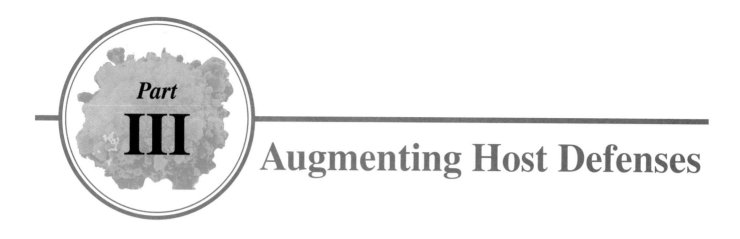

Part III

Augmenting Host Defenses

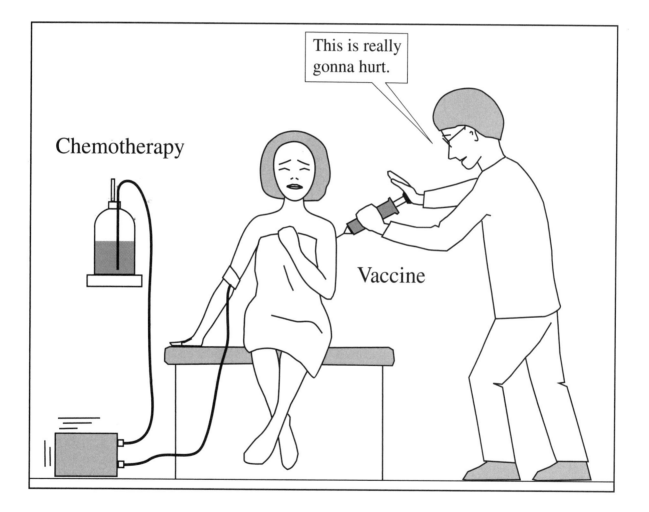

In Lecture 2 we reviewed the potent, multilayered defenses that have evolved to protect humans from viral infections. Then, in subsequent lectures, we surveyed some of the sly schemes viruses have come up with to evade these defenses long enough to reproduce and spread. In many instances, our defenses are adequate to prevent serious disease.

However, in other cases (e.g., during an HIV-1 or herpes simplex virus infection), we clearly could use some "outside" help. In the remaining lectures, we will examine the two approaches which have been most successful in augmenting host defenses: vaccinations and the use of antiviral drugs.

Lecture

8

Vaccines

R E V I E W

In the last lecture, we talked about emerging viruses: viruses which have only recently become apparent. Although it is possible that a brand new virus may evolve during our lifetimes, it is very unlikely. After all, every virus must learn to solve the problems of reproduction, spread, and evasion of host defenses, and this usually takes a very long time. No, it is much more probable that viruses will "emerge" from existing viruses. This can happen when technological advances make it possible to detect existing human viruses; when changes in human lifestyles allow existing animal viruses to take hold in the human population; when humans "turn over rocks," exposing themselves to viruses which were formerly living in harmony with their animal hosts, and when existing viruses mutate to produce closely related viruses that can infect humans.

Hypervirulent viruses like Ebola and hanta have been billed as "killer viruses of the future" by the press. However, these viruses are not especially dangerous to the human population, mainly because humans are not natural hosts for these viruses. In fact, one could hypothesize that a really dangerous virus (one that I called the Andromeda Strain) would be very different from the Ebola or the hanta virus and would have the following properties: The Andromeda Strain would have humans as a natural host; it would be easily spread by casual contact; it would kill humans only after a long, relatively asymptomatic period during which it could spread to other humans; and it would arise from a present-day virus that has a high mutation rate.

We discussed two contemporary viruses, influenza A and HIV-1, which have many, but not all, of the properties we would predict for the Andromeda Strain. Influenza A virus mutates rapidly, has humans as one of its natural hosts, and is spread by casual contact. However, when a deadly strain of influenza virus emerged in 1918, it killed its hosts so rapidly that only a small fraction of the human population was infected before it burned itself out. HIV-1 is a virus that also mutates rapidly, and which kills essentially everyone it infects, usually after a decade-long infectious period. However, HIV-1 is not spread by casual contact, so most people can avoid HIV-1 by not having intimate physical contact with infected individuals.

Although neither HIV-1 nor influenza A qualifies as the Andromeda Strain, they come frighteningly close. From our experiences with these two near misses, it is clear that we must take emerging viruses very seriously, and that virologists must try to evaluate the pathogenic potential of such viruses early after their emergence.

VACCINES

Vaccines are incredibly powerful weapons for controlling infectious diseases. Indeed, smallpox virus, which once killed millions, has been eradicated from the earth through the use of a vaccine. Although a number of effective antiviral vaccines do exist, there are many viral infections for which no good vaccine has been produced. For example, thus far all attempts to make a vaccine that would protect against an HIV-1 infection have failed. Indeed, because HIV-1 is a showcase for the problems that can arise in producing a useful vaccine, we will use HIV-1 as our "model" virus in this lecture. As we discuss strategies currently used to make vaccines, we will evaluate whether any of these approaches might be suitable for producing a safe and effective AIDS vaccine. I think you'll come to appreciate that this is really a knotty problem—one that may turn out not to have a solution.

Generating Memory Cells

The goal of any vaccination is to generate memory B and T cells, for these are the cells that can provide powerful protection if the vaccinated person is subsequently exposed to the "real thing." To create a vaccine that will produce memory cells, immunologists employ strategies that are similar to those used by the army in its war games. In these exercises, the generals prepare their troops for battle by presenting them with as realistic a look at war as is possible—without putting them in great danger. Likewise, vaccine developers attempt to give the immune system the most realistic possible "look" at an invader (e.g., HIV-1) without endangering the vaccine recipient. In effect, the immunologist's creed in devising a vaccine is: maximum realism with minimum danger.

To understand exactly what the immune system must "see" in order to generate memory cells, let's review how memory cells are made in response to a viral infection. The important point here is that generating memory killer T cells requires a different "look" at the invader than does generating memory helper T or B cells.

When a person is exposed to a virus for the first time, the innate system does its best to repulse the attack. If the innate system is unable to deal quickly with the invasion, a vigorous inflammatory reaction is generated in which battle cytokines are produced. Dendritic cells that have phagocytosed viruses or viral debris at the battle site are activated by these battle cytokines, and travel with their cargo of viral antigens to nearby lymph nodes. During this trip, dendritic cells cease being antigen collecting cells and become antigen presenting cells. Once in the lymph nodes, dendritic cells use their class II MHC molecules as billboards to display the viral antigens they have acquired. As helper T cells circulate through lymph nodes, they inspect these billboards. If their T cell receptors recognize the viral antigens, and if they receive the appropriate co-stimulation, these virus-specific helper T cells will be activated. After a period of proliferation, the activated helper T cells assist in the activation of B cells whose receptors recognize virus particles or viral debris that has been carried to the lymph node by the lymph. These virus-specific B cells then proliferate, and can undergo class switching and somatic hypermutation. After all this action, some of the helper T cells and B cells are "chosen" by a still-mysterious process to become memory cells. So for memory B and helper T cells to be generated, a B cell's receptors must "see" either viruses or viral debris, and a helper T cell's receptors must "see" fragments of viral proteins that have been taken up and displayed by dendritic cells.

Killer T cells must also be activated in order to produce memory cells. However, to be activated by a viral infection, a killer T cell must recognize a fragment of a viral protein displayed by class I MHC molecules on the surface of an antigen presenting cell. This occurs most efficiently when the virus actually infects a dendritic cell out in the tissues and rides with this antigen presenting cell to a nearby lymph node. Consequently, there is an important difference in the requirements for generating memory B and helper T cells vs. memory killer T cells: The immune system can produce memory B and helper T cells even in response to viral debris. In contrast, to efficiently generate memory killer T cells, antigen presenting cells must be infected by a virus.

With this as background, we can now begin to examine the vaccine strategies used to produce memory cells. As we discuss each type of vaccine, we will pay special attention to the kinds of memory cells the vaccine generates, and how safe each vaccine is likely to be.

Noninfectious Vaccines

One strategy used for memory cell production involves vaccination with an agent that cannot infect the vaccine recipient. An example of such a "noninfectious" vaccine is the killed polio virus vaccine devised by Dr. Jonas Salk. To produce this vaccine, Salk and his coworkers first grew large quantities of polio virus in the laboratory by infecting monkey cells growing in petri dishes. Then they treated this virus with chemicals such as formaldehyde that "glue" viral proteins together. In some ways, this treatment is similar to a policeman applying the

"Boston Boot" to a car with too many parking tickets. The car looks just fine, but it isn't going anywhere because its wheels won't turn. Likewise, to the immune system, Salk's killed virus looks just fine, but it can't cause an infection, because it has been disabled by the formaldehyde treatment. Noninfectious vaccines are now quite common. For example, the flu vaccine we get each year is a killed virus vaccine.

One important characteristic of vaccines made from "killed" viruses is that although the chemical treatment will certainly disable most of the viruses, it is impossible to guarantee that it will kill them all. This is not a big problem if the disease in question is spread by casual contact and if a relatively large fraction of the population is at risk of infection—as was the case with polio in the 1950s. However, for a virus like HIV-1, which in most cases can be avoided, a killed virus vaccine that has even a remote possibility of causing disease would be unacceptable.

Another strategy for making noninfectious vaccines is to use genetic engineering to produce one or a few viral proteins in the laboratory, and to use these proteins as a vaccine. Such a vaccine is called a "subunit" vaccine, and this technique is currently used to make a very effective vaccine against hepatitis B virus. Subunit vaccines have the great advantage that, because the genes for only one or a few viral proteins are present, there is no possibility that the vaccination will result in an infection with the virus from which the genes were taken. On the other hand, the limited number of viral proteins included in a subunit vaccine gives the immune system fewer targets to focus on, and this paucity of targets can be problematic when vaccinating against viruses that have high mutation rates.

Now here is an important question: What kind of memory cells do you expect a vaccine made from a killed virus to produce? Because they can be activated by microbes that don't infect human cells, memory B and helper T cells will be made in response to a noninfectious vaccine. As a result, this type of vaccine can elicit the production of protective antibodies. But what about killer T cells that can destroy virus-infected cells? Will a noninfectious vaccine elicit the production of killer T cells? The answer is that few, if any, killer T cells will be produced in response to a noninfectious vaccine. To efficiently activate killer T cells, the agent used for the vaccination must infect antigen presenting cells—and by definition, a noninfectious vaccine can't do that.

Whether or not the lack of killer T cells will be a problem in controlling a viral infection is hard to predict. For example, the Salk polio vaccine works very well, so clearly a noninfectious vaccine that causes the production

of virus-specific antibodies can defend against a polio virus infection. On the other hand, killed virus vaccines for measles and mumps were real duds. So it just depends on the virus.

There is a strong feeling among immunologists that a killer T cell response will be essential to resist an HIV-1 infection. Despite this belief, a noninfectious, subunit vaccine against HIV-1 is currently being tested on about 2,500 volunteer intravenous drug users in methadone centers in Bangkok, Thailand. The rationale for using these test subjects is that without vaccination, about 5% of the volunteers will be infected with HIV-1 each year. As a result, it will take a relatively short time to determine whether the vaccine has had an effect. A similar vaccine is being tested on about 5,000 Americans who are at risk for contracting an HIV-1 infection. The results of these two trials should be very interesting, because they may demonstrate that it is not necessary for an AIDS vaccine to produce memory killer T cells. Memory B cells that make large quantities of neutralizing antibodies may be enough.

A major concern with this and other AIDS vaccines is that HIV-1 may mutate so quickly that memory cells produced by the vaccination may not be appropriate to defend against the mutated virus. This would be somewhat similar to this year's influenza vaccine not being able to protect against next year's flu strain. The results of the Bangkok and American trials may also shed some light on this potential problem.

Attenuated Virus Vaccines

The famous Sabin polio vaccine was produced using another strategy. To make this "attenuated" vaccine, Dr. Sabin grew the polio virus in monkey kidney cells rather than in human nerve cells. For reasons that are not well understood, growing a virus in the "wrong" host can produce mutations in the virus that weaken it. In this case, Sabin's experiments resulted in three strains of polio virus that could infect a vaccine recipient, but which were so weak (attenuated) that they could not cause disease in healthy people. These three strains were then combined to make the Sabin vaccine. A similar strategy has been used to produce the attenuated mumps, rubella, and measles vaccines that are in common use today.

Preparation of an attenuated virus vaccine involves a bit of magic, because attenuated viruses must walk a fine line: They must be strong enough to produce a vigorous immune response, yet weak enough not to cause disease. To determine whether the attenuation strategy has worked, the vaccine can be tested on animals, assuming an appropriate animal can be found. But ultimately, the vaccine must be tested on human "volunteers." Interestingly, by the time Sabin was ready to test his vaccine, most Americans had already been vaccinated with the Salk vaccine, and of course, it wouldn't make sense to test his vaccine on Americans who had already been vaccinated against polio. So Sabin had to look elsewhere. And where do you think Sabin went to find his volunteers? To Russia! No, I'm not kidding you. During the Cold War, Sabin tested his vaccine in Russia—a country with a well-developed health care system, whose people had not yet been vaccinated against polio.

Vaccines made from attenuated (weakened) microbes have an important advantage: they usually provide lifelong immunity. This is in contrast to noninfectious vaccines which frequently require periodic booster vaccinations to maintain protective antibody levels. Although it is easy to imagine how a real infection, albeit attenuated, might give "better" immunity than vaccination with a dead virus, the immunological basis for this difference is not completely clear—probably because immunologists still do not fully understand how long-term immunological memory is maintained.

Before being eradicated by a vaccine recipient's immune system, an attenuated virus can infect antigen presenting cells. So attenuated vaccines can produce memory killer T cells. Of course, the fact that an attenuated virus actually infects the recipient raises several safety issues. First, although a healthy immune system usually will wipe out an attenuated virus before it can cause serious disease, this may not be the case for an individual whose immune system has been weakened. In fact, if a healthy person who has just been inoculated with an attenuated virus vaccine passes the virus on to someone who is immunosuppressed (e.g., due to cancer chemotherapy), there is a chance that the immunosuppressed person's immune system will be too weak to fight off the attenuated virus. Also, because the attenuated virus will multiply to some extent in a healthy vaccine recipient, there is a slight possibility that the crippled virus will mutate, and that it will again become strong enough to cause disease. The probability that such a mutation will occur is low, but even the Sabin attenuated vaccine has caused a few cases of polio in healthy recipients due to mutations that restored the strength of one of the three attenuated viruses used in this vaccine. Because the AIDS virus has an extremely high mutation rate, this could be a real concern if HIV-1 were attenuated and used as a vaccine.

Carrier Vaccines

To efficiently produce memory killer T cells, a vaccine must be able to infect antigen presenting cells. However, safety concerns dictate that an infectious, attenuated AIDS virus would be unacceptable because of the risk of mutation. To try to resolve this dilemma, immunologists are experimenting with new approaches to vaccine design.

One new strategy involves using a virus or a bacterium that doesn't cause disease (e.g., the vaccinia virus used for decades to vaccinate against smallpox) to "carry" one or more HIV-1 genes into antigen presenting cells. In this way, carrier-infected cells could be tricked into producing several HIV-1 proteins in addition to the carrier's own proteins. Consequently, killer T cells that recognize these HIV-1 proteins could be activated. Most importantly, although a carrier vaccine should produce memory killer T cells, there is no chance that the vaccine will cause a real infection—because only a few HIV-1 genes are "carried" by the vaccine.

On the surface, it would seem that a carrier virus vaccine would be just the ticket for protecting the general population against HIV-1, and vaccines that use various viruses and bacteria as carriers are currently being tested. There are, however, potential problems with these vaccines. First, as with an attenuated virus, it is possible for the recipient of a carrier vaccine (e.g., one which uses vaccinia virus as a carrier) to pass that carrier on to others. Although this has the advantage of spreading the vaccination around, it also has the risk of infecting immunocompromised individuals whose immune systems may not be able to subdue the carrier infection.

Second, if a person has previously been exposed to the carrier, that person will have memory B and T cells which will protect against a future carrier infection. So if that same carrier is then used as a vaccine, the already alerted immune system will usually shut down the carrier infection before sufficient proteins are made to cause an effective vaccination. As a result, the carrier must be carefully chosen to be one to which most vaccine recipients are unlikely to have been exposed. For example, since about 1970, vaccinia virus has not been used to vaccinate against smallpox—so vaccinia virus should work fine as a carrier for people under thirty. However, for us older folks, who were vaccinated with vaccinia as children, a different carrier would have to be used.

A third concern about carrier vaccines is that the vaccination may need to be repeated (boosted) in order to be effective. Unfortunately, boosting doesn't appear to work with carrier vaccines. This is because vaccination with a carrier virus usually produces enough memory B and T cells to "protect against" a second, booster vaccination. Of course, one way to overcome this problem is to use a different carrier for the booster vaccination. Another is to use a carrier vaccine for the initial vaccination, and a subunit vaccine to boost the number of memory B cells that are made. This strategy (called "prime-boost") is currently being used in an HIV-1 vaccine trial involving over 400 test subjects. Recent results indicate that most of the people who received the canarypox (a relative of smallpox virus which normally does not infect humans) carrier vaccine plus a subunit vaccine boost developed antibodies against the strain of HIV-1 used to prepare the vaccines. In addition, about 30% of the recipients developed killer T cells that could kill HIV-1-infected cells in the laboratory. However, much more testing will be needed to determine whether these immune responses are strong enough to be protective.

DNA Vaccines

Several years ago, immunologists made a remarkable discovery. When a piece of DNA that included an influenza virus gene was injected into the muscles of mice, many of these mice were protected against a subsequent influenza infection that normally would have been lethal. Importantly, it was later shown that DNA injections could produce not only memory B cells, but memory killer T cells as well. The immunological basis for this rather amazing effect is still not fully understood, but it is suspected that dendritic cells in the muscle tissues took up the DNA, produced the protein specified by the influenza gene, and presented fragments of the flu protein to activate killer T cells.

This totally unexpected finding led to the suggestion that naked DNA, containing genes from disease-causing microbes, might be used to vaccinate humans against these pathogens. Because a "DNA vaccine" would be made from genes that encode only one or a few viral proteins, there would be no danger that the vaccination could cause the disease it was designed to protect against. Also, because memory killer T cells would be produced, DNA vaccines could be used against viruses that require killer T cells for protection. DNA vaccines that include genes for HIV-1 proteins are currently being tested in animals.

Vaccine Testing

Developing a new vaccine is a long, complicated, and costly process. First, a vaccine is tested in an appropriate animal model to be sure that, at least in an animal, the vaccine is safe and provides protection against the infectious agent. To be "appropriate," the animal should be susceptible to infection by the agent against which the vaccine is directed, and should develop the same disease symptoms as humans do. This is an important first step, because some vaccine strategies that look promising in a test tube turn out not to work in an animal.

For some viruses, a good animal model is available, but for HIV-1, this is not the case. Although chimpanzees can be infected with the common strains of HIV-1, they do not develop AIDS symptoms, indicating that there are fundamental differences between chimps and humans as far as HIV-1 infections are concerned. On the other hand, macaque monkeys develop AIDS-like symptoms when infected with SIV, a monkey virus which is a close relative of HIV-1. So a vaccine against SIV can be tested in macaques. However, there is no guarantee that an HIV-1 vaccine made using the same strategy would protect humans against HIV-1.

If, using the best available animal model, a vaccine is shown to be safe and effective in animals, the next step is to carry out a Phase I human trial. This is an important test, of course, because a vaccine that works in animals may be ineffective or may have serious side effects in humans. For a Phase I trial, a small number of volunteers (generally fewer than 100) are recruited. These are usually healthy people who have a vested interested in vaccine development—for example, individuals who are likely to be exposed to the virus at some time in the future. The goal of a Phase I trial is to determine vaccine dosages that produce levels of memory B or T cells that are likely to be protective, and to evaluate side effects that may result when humans are vaccinated at these dosages.

If the results of a Phase I trial indicate that the vaccine preparation is safe, and that memory cells are produced, a Phase II trial can be undertaken. A Phase II trial involves a larger number of volunteers (usually several hundred) who are expected to have varying risks of being exposed to the virus. Although the main focus of both Phase I and Phase II trials is on safety, a Phase II trial may also yield information on the effectiveness of the vaccine. Phase II trials usually are "controlled" studies in which some of the volunteers are given either a non-protective (placebo) vaccination, or an existing vaccine for comparison.

If the results of a Phase II trial look promising, the vaccine can be tested on a large number of individuals at high risk for infection. The goal of a Phase III trial is to evaluate the effectiveness of the vaccine, and to test for infrequently occurring side effects. Phase III trials are almost always "double blind" in that neither the patient nor the physician knows whether the "vaccinations" contain the real vaccine or a control (e.g., a placebo) vaccine. This insures that patients don't somehow resist the attack of the virus better, just because they think they have been vaccinated against it (the "placebo effect"), and that the physician gives the same care to both vaccinated and control-vaccinated individuals. Only when the trial is complete are the codes broken and the results evaluated.

If a Phase III trial shows that the vaccine has relatively minor side effects, and is effective in protecting most vaccine recipients (usually 90% or greater), it can be licensed by the Food and Drug Administration for general use. For an average vaccine, it takes several years to carry out three phases of human trials. However, to test whether an AIDS vaccine is protective, the trial period may have to be longer, because HIV-1 can remain in a latent state for years, and during this time, even the most sensitive techniques may be unable to detect its presence.

Vaccines as Treatments

We tend to think of a vaccination as something one gets for protection against a possible future exposure to a disease-causing agent. However, in some cases, vaccines can also be used to treat a person who has already contracted a disease. The most common example of such a vaccine is the rabies vaccine which usually is given after a person has been bitten by a rabid animal. Post-exposure vaccines like the rabies vaccine work well if the vaccine can activate the immune system long before the invader has a chance to take over. In the case of rabies, the virus reproduces so slowly that it may take a month or more before disease symptoms appear.

A post-exposure vaccine for AIDS has recently been developed by Dr. Jonas Salk and his co-workers. Here the hope is that the immune system of the vaccinated person will be stimulated to better resist the virus, so that the onset of AIDS symptoms will be delayed, and the period between contracting the disease and death will be lengthened. In this case, the post-exposure vaccine would be used as an AIDS treatment (like AZT or protease inhibitors), not as a cure. Because this is a killed virus vaccine, it is unlikely that it could be used safely to protect uninfected people against HIV-1. However, once a person has been infected, the slight possibility that the vaccine might contain some residual live virus is no longer an issue. Phase I and II trials of the vaccine suggest that it is safe (i.e., there are no unexpected side effects), and that it may slow the progression of the disease. A Phase III trial involving over 2,000 infected Americans is now in progress to better assess the benefits of this therapeutic vaccine.

Prospects for an Effective AIDS Vaccine

As I'm sure you realize, the prospects for developing a vaccine that will protect the general population against an HIV-1 infection are not that great. The virus' ability to establish a latent infection that is invisible to the immune system, and its high mutation rate make it an elusive target. Although there are indications that some AIDS vaccines can activate the immune system against the same strain of HIV-1 used to create the vaccine, it still has not been shown that these vaccines will protect against the mutated strains that arise during a real infection.

Perhaps the biggest obstacle to producing an effective AIDS vaccine is that immunologists really don't know what type of immune response is required to protect against HIV-1. For most other diseases for which effective vaccines have been produced, there are individuals who have been infected by the microbe and whose immune systems have rid their bodies of the attackers. For example, there are many people who have been infected with smallpox virus, and whose immune systems have fought off the virus. By examining these survivors, immunologists can discover whether they resisted the disease because of an antibody response, a killer T cell response, or both. In fact, immunologists can even determine which viral proteins the protective immune response is directed against—an insight that can be very useful in designing vaccines.

With AIDS, there are individuals who seem to be able to resist HIV-1 infection in settings in which they normally would be expected to contract the disease. There are also "slow progressors" who have contracted HIV-1, but who have shown no symptoms of the disease, even after many years. However there is no documented case of someone who has had a full-blown HIV-1 infection and whose immune system has fought off and destroyed the virus. In fact, because no one seems to have survived a "natural" HIV-1 infection, there is the underlying fear that vaccination against the AIDS virus may not be possible: The human immune system simply may not be capable of mounting a successful defense against this particular virus. However, the hope remains that some clever immunologist will come up with a vaccine that can prepare the immune system to resist an HIV-1 attack—perhaps in ways that exposure to the virus itself cannot do.

Although HIV-1 poses difficult problems for vaccine development, it is certainly not the only virus against which there is no effective vaccine. Indeed, good vaccines exist for only four of the twelve viruses in our Bug Parade. For example, herpes simplex virus infects about one third of the world's population, yet there is no vaccine that can protect against a herpes simplex infection. Rotavirus kills nearly a million children worldwide each year, yet the only rotavirus vaccine in production was withdrawn from the market in 1999 because it killed some of the children who received the vaccination.

So devising a safe and effective vaccine is not a simple project. To approach the problem of treating or preventing viral infections in another way, virologists have been trying to discover drugs that can interfere with critical steps in the reproductive programs of some viruses. However, as we will discuss in our next lecture, producing useful antiviral drugs isn't easy either.

Antiviral Drugs

Vaccines have been extremely useful in protecting against viral attacks. In contrast, once a person has been infected, the drugs available to treat a viral infection are very limited. At first glance, this doesn't seem to make much sense. After all, antibiotics are used to treat a wide range of bacterial infections. So why are there no broad-spectrum antiviral drugs with the same capability?

Bacteria tend to be "free-living" organisms which have evolved ways of doing business that are sometimes very different from the ways things are done in human cells. For example, the walls of human cells and bacterial cells are assembled from very different materials. As a result, an antibiotic such as penicillin, which disrupts the synthesis of a common bacterial cell wall component, can kill a wide range of bacteria without damaging human cells.

In contrast, viruses are almost totally dependent on the biochemical machinery of human cells for their reproduction. Consequently, it is very difficult to discover drugs that will kill viruses but not human cells. Indeed, most antiviral drugs have significant side effects that make them unsuitable for long-term use in chronic viral infections. Also, because individual viruses have solved their problems in so many different ways, it is difficult to imagine the possibility of a broad-spectrum antiviral drug. Indeed, those lifestyle features that are uniquely viral generally are common to only a small number of viruses.

The lack of broad-spectrum antivirals is a real problem. Many bacterial infections cause similar symptoms, but by treating these infections with broad-spectrum antibiotics, the exact identity of the infecting bacterium usually need not be determined. In contrast, the lack of broad-spectrum antivirals means that the invading virus must frequently be identified before treatment can begin. And in many cases, this identification takes so long that by the time the appropriate antiviral drug can be selected, an acute viral infection will already have run its course—or a chronic or latent infection will already have been established.

Broad-spectrum antivirals aside, it is even difficult to produce an antiviral drug that is effective against a single virus. To rationally design an antiviral drug, virologists must first identify a target that is uniquely viral. This usually requires a detailed understanding of the virus' reproductive program. However, this information is only available for a small number of viruses, so many of the current treatments were discovered by testing "every drug on the shelf" for antiviral activity.

As a first step in determining the efficacy of a potential antiviral compound, virologists would like to test virus-infected human cells in the lab to determine whether treatment with the antiviral reduces virus production while sparing uninfected cells. However, for several of the most important viruses (e.g., hepatitis B, hepatitis C, and the human papilloma virus), good systems do not exist for growing the virus in the lab. Further, if a compound works well on virus-infected cells, virologists would next like to test whether the antiviral will be effective in an animal. Such experiments are vital to determine whether the drug will reach the right part of the body to work its magic, and whether it can be maintained in the animal at an effective concentration without being toxic. Unfortunately, good animal models for viral infection do not exist for some very important viruses (e.g., HIV-1).

Finally, drug development is insanely expensive and time consuming: It usually costs at least 100 million dollars and takes five to ten years to test and license a new antiviral drug for general use. This fact is often overlooked by consumers who complain that prescription drugs cost too much.

If you don't count "variations on a theme," fewer than a dozen antiviral drugs have been approved for general use by the FDA. Because there are so few antivirals to discuss, this will be a short lecture!

Targets for Antiviral Drugs

The goal of an antiviral drug is to interfere with some step during a viral infection that is uniquely viral. These targets can be arranged into five main groups, which represent stages in the virus life cycle: entry, uncoating, genome replication, virion assembly, and exit. In this lecture, we will evaluate how successful virologists have been in developing drugs that target each of these stages, and we will focus our discussion on what might be called "prototype" drugs. Pharmaceutical companies are constantly coming up with refinements (analogs) of these prototypes that act in the same way, but which have fewer side effects, are better absorbed, have longer half-lives in the body, etc. Because our main interest here is understanding how various drugs interfere with virus lifestyles, we will concentrate on the mechanisms of drug action, and won't worry about trying to keep up with the latest analogs.

Viral Entry

Because neutralizing antibodies are able to prevent entry of some viruses into their target cells, it would seem that drugs also could be discovered or designed that could block viral entry. Thus far, however, there is no commercial drug which can do this. Part of the problem with this approach is that some viruses (e.g., herpes simplex and HIV-1) have alternative cellular receptors which they can use to gain entry. So a combination of drugs would be needed which could block the interaction of a virus with all of its possible receptors.

Another difficulty in designing therapeutic drugs which prevent viral entry is that it involves a "numbers game" which is heavily weighted against the drug. A viral capsid or envelope contains many copies of the proteins that bind to cellular receptors, and the surface of a cell usually has thousands of these receptors. So to block entry, a drug would have to interact with a significant fraction of the viral proteins or cellular receptors—and this is a tall order. In some cases, antibodies can win this

numbers game because a huge number of antibody molecules can be produced and transported to the site of infection. Moreover, antibody molecules are extremely stable, with half-lives that usually are measured in days or weeks. In contrast, high concentrations of most therapeutic drugs are toxic, and many drugs persist within the body for only a few hours before they are inactivated by binding to proteins in the blood (e.g., albumin), excreted in the urine, or degraded in the liver.

Viral Uncoating

An excellent example of a therapeutic drug that interferes with viral uncoating is **amantadine.** This drug, which can be used to treat influenza A infections, is now over thirty years old. It inhibits the activity of a viral protein, M2, which is required for uncoating. The envelope of influenza A virus contains a relatively small number of M2 proteins (fewer than seventy). Moreover, multiple M2 proteins must fit together properly to form a channel through which protons can enter the interior of the virus to facilitate uncoating. As a result, relatively few M2 molecules need to be "compromised" by the drug to keep the ion channels from working properly and to trap the virus within its coat.

Unfortunately, even this highly specific antiviral drug has its problems. First, the influenza virus mutation rate is so high that essentially every virus produced is different from the original infecting virus. As a result of this antigenic drift, influenza mutants arise whose M2 proteins no longer interact with amantadine. Second, to be effective in decreasing the severity of an influenza attack, amantadine must be administered during the first two days after infection—and most people don't even realize they have the flu for the first day or two. Moreover, because other microbes cause flu-like symptoms, it is difficult for a physician to be sure that amantadine is the drug of choice. For example, influenza B virus causes early symptoms similar to those caused by influenza A virus, yet influenza B virus, which has no M2 protein, is unaffected by amantadine. Consequently, amantadine is usually prescribed to protect against an anticipated influenza A infection (about 60% effective) rather than as a treatment.

Ideally, since influenza virus infects cells of the respiratory tract, one would like to bathe these cells in the drug. However, this is not very practical, so amantadine is administered orally. As a result, about 90% of the drug is excreted in the urine, and the high doses required to achieve a therapeutic effect on cells in the airways can lead to serious side effects involving the central nervous system.

Viral Genome Replication

Although viruses rely on cellular machinery for their replication, all viruses replicate their genomes using strategies which are at least somewhat different from that used for the replication of cellular DNA. These idiosyncracies in viral genome replication represent excellent targets for antiviral therapies.

AZT is a drug that takes advantage of a unique feature of viral replication and is widely used to treat HIV-1 infections. When HIV-1 replicates, it employs its reverse transcriptase enzyme to make a complementary DNA copy of the virus' RNA genome. During this operation, DNA building blocks (the nucleotides) are strung together according to the "recipe" contained within the single strand of HIV-1 RNA. Each nucleotide has a "plug" (the 5' tri-phosphate) and a "socket" (the 3' hydroxyl), and new nucleotides are added by plugging them into the socket of the nucleotide that is at the end of the growing DNA chain. AZT is converted by cellular enzymes into a "fake nucleotide" which has a plug, but no socket. Consequently, when AZT is incorporated into DNA, there is no socket for the next nucleotide to plug into, and the DNA chain terminates. The result of this premature termination is a viral cDNA that cannot be integrated into the cellular DNA of the infected cell—and without integration, the viral infection is aborted.

Unfortunately, although AZT can interfere with HIV-1 replication, it also can be incorporated into cellular DNA in proliferating human cells. When this happens, these cells can be killed. Fortunately, the HIV-1 reverse transcriptase has a higher affinity for the AZT nucleotide than do most cellular polymerases. Consequently, there is a "therapeutic window" of AZT concentrations that favors incorporation into viral vs. human DNA. Nevertheless, there are many cells in the body which proliferate almost continuously (e.g., cells that line the digestive tract and blood cells in the marrow), and these cells can be killed by the AZT treatment, producing substantial side effects. This problem is exacerbated by the fact that the half-life of AZT in a human is only about one hour because the drug is rapidly degraded by the liver. So keeping the AZT concentration high enough to interfere with viral replication without producing serious side effects is a challenge. In addition, because it is difficult to maintain a therapeutic level of AZT, the virus continues to replicate when concentrations wane—and mutant viruses arise whose reverse transcriptase enzymes will no longer incorporate AZT into their cDNAs. For these reasons, AZT used alone is only moderately useful in treating a chronic HIV-1 infection. In contrast, a short course of AZT treatment is usually quite effective in reducing the probability that an infected mother will transmit the AIDS virus to her child at birth.

Acyclovir is another nucleoside analog that can act as a DNA chain terminator. The real beauty of acyclovir, however, is that the conversion from nucleoside analog to fake nucleotide is not carried out efficiently by cellular enzymes, so uninfected cells are rarely damaged by this drug. In contrast, both herpes simplex virus and varicella-zoster virus (which causes chickenpox and shingles) encode the enzyme, thymidine kinase, which, together with cellular enzymes, efficiently converts acyclovir into a fake nucleotide. As a result of this selectivity, acyclovir is very useful for treating herpes simplex and varicella-zoster infections.

Ribavirin is a nucleoside analog that does not function by terminating nucleic acid synthesis. This drug has both a plug and a socket. Indeed, until very recently, the mode of action of this antiviral drug was quite mysterious. Now it has been discovered that ribavirin acts by increasing the rate at which viral RNA polymerase enzymes introduce mutations into the RNA molecules they make. Here's how this works.

As we have discussed, human RNA viruses have a high mutation rate because of the error-prone nature of their RNA polymerases. This high rate of mutation is a great benefit to these viruses, because it makes it more likely that the viral population created during an infection will include mutants which can resist newly evolving host defenses. However, this high mutation rate can also be dangerous to the virus: If the rate is too high, many newly made viruses will be so badly mutated that they cannot reproduce. So RNA viruses must walk a fine line between mutating too slowly and falling prey to evolving host defenses, and mutating so rapidly that they become nonfunctional. It has been proposed that the best solution to this problem is for a virus to mutate to the max: to "adjust" its mutation rate to be slightly below the point at which the errors introduced begin to seriously decrease the viability of the virus. What this means is that if one could treat virus-infected cells with a drug which would substantially increase the already high viral mutation rate, the RNA viruses replicating in those cells might be forced into "error catastrophe." And that's just what ribavirin does.

When ribavirin is taken up by a cell, it is modified (phosphorylated) by host enzymes to become a fake G (guanosine) RNA nucleotide. Once ribavirin has been incorporated into the position normally occupied by a real G nucleotide, it can be copied along with the rest of the RNA molecule when the virus replicates. However, whereas the real G nucleotide templates the addition of a

C (cytidine) nucleotide to the growing, complementary RNA chain, ribavirin templates the addition of C or U (uridine) nucleotides with roughly equal efficiencies. And introducing a U instead of a C results in a mutation. If the concentration of ribavirin within the cell is high enough, the virus may be driven into an error catastrophe in which most of the newly made RNA genomes are nonfunctional. Reaching the concentration required for lethal mutagenesis is made easier by the fact that ribavirin also inhibits a cellular enzyme required to produce real G nucleotides. This decreases the pool of real G nucleotides within the cell, and increases the probability that the fake G will be incorporated into growing viral genomes.

Although ribavirin has been used to treat a number of different viral infections and is frequently cited as a broad-spectrum antiviral, it has only been conclusively demonstrated to be efficacious in treating hepatitis C infections (in conjunction with interferon) and respiratory syncytial virus infections in infants. It is hypothesized that ribavirin's lack of clear broad-spectrum activity may result from the differing efficiencies with which various viral RNA polymerases incorporate the fake bases into their genomes. Alternatively, different viruses may be more or less susceptible to the mutagenic activity of ribavirin because they have chosen to "walk" at different distances from the error catastrophe "precipice."

In addition to drugs like AZT and acyclovir, which are incorporated into growing viral DNA chains, there are also drugs which interfere with synthesis of viral DNA by binding to the viral polymerase molecule. One such drug, **foscarnet,** binds to the polymerase enzymes of hepatitis B virus, HIV-1, and certain herpes viruses (e.g., herpes simplex, human cytomegalovirus, vericella-zoster virus). It is interesting that even though the polymerases of these viruses are very different, this drug can bind to them all. So foscarnet is about as close to a broad-spectrum antiviral as has been discovered. However, foscarnet is a highly charged molecule, so transport across the cell membrane is inefficient. As a result, therapeutic doses are usually so high, and the side effects so serious (kidney toxicity) that forscarnet is generally a "last resort" drug, which is mainly used in immediate, life-threatening illnesses.

Another non-nucleoside polymerase inhibitor is **nevirapine.** This drug binds to the HIV-1 reverse transcriptase at a site just next to the active site of the enzyme, creating a distortion that interferes with the polymerase's ability to produce viral cDNA. There is now a whole family of non-nucleoside, reverse transcriptase inhibitors like nevirapine, and these drugs are used extensively in treating HIV-1 infections.

We have already discussed the **type I interferons** (interferon alpha and beta) which are "natural" inhibitors of viral replication. Unfortunately, many viruses have evolved clever defenses that mitigate against the effects of interferon. Consequently, there are only a few viruses for which interferon treatments have been successful. It is believed that when interferon works, it is because the drug reduces production of new viruses to a point where the immune system of the patient can take control of the infection. Interferon alpha has been used to treat hepatitis B infections, with roughly 30% of those treated showing a long-term response. The results for interferon alpha treatment of hepatitis C infections have been less positive, with only certain strains of hepatitis C showing a lasting response. These are presumably hepatitis C strains which have not invested heavily in an anti-interferon defense.

Virus Assembly

During the final stages of assembly of HIV-1 virions, a viral enzyme (the HIV protease) cuts a large Gag-Pol precursor protein into pieces to yield viral structural proteins and enzymes. This protease activity is unique to the virus (i.e., there is no cellular enzyme that can do this cutting), so it has become an attractive target for anti-HIV-1 drugs. One representative protease inhibitor, **indinavir,** mimics a cleavage site that the HIV protease recognizes in the Gag-Pol protein. At sufficiently high concentrations, indinavir can "distract" the viral protease and keep it from making the cut in the Gag-Pol protein required for viral assembly.

Other viruses, for example herpes viruses, encode proteases that are required for viral assembly. So far, however, the HIV-1 protease inhibitors are the only drugs on the market which target this step in the virus life cycle.

Virus Exit Inhibitors

When influenza virus exits a cell it has infected, it must solve a problem. Because the viral envelope contains both hemagglutinin and hemagglutinin receptors, there is a possibility that influenza viruses will bind to each other as they exit, producing noninfectious clumps of virus. In addition, the hemagglutinin proteins on the exiting viruses can bind to receptors on the cell they are leaving, trapping the virus on the surface of a cell that has already been infected. As we discussed in Lecture 3, influenza virus deals with this exit problem by producing a "razor-like" protein, the viral neuraminidase. This protein is inserted into the cell membrane, where it "shaves" the cellular receptors, removing the sialic acid residues to which the hemagglutinin proteins bind. Because this neuraminidase activity is so important for viral spread, the viral neuraminidase protein is an excellent target for anti-influenza drugs.

Relenza is a drug which is inhaled, and the drug, **Tamiflu,** is taken orally. Both are neuraminidase inhibitors, and both are effective against both type A and type B influenza. These drugs are sialic acid mimics to which the neuraminidase enzyme binds much more tightly than to real sialic acid residues. As a result, the neuraminidase enzyme tries to "shave" the drug rather than the cell. If these drugs are used within the first two days after infection, they can shorten the duration of symptoms by a day or two. Of course, as we discussed earlier, it is difficult to self-diagnose an influenza infection within the first forty-eight hours, so these drugs are generally used for people living in close proximity after one or more of them has already been diagnosed with an influenza infection (e.g., during a flu outbreak in a nursing home).

Combination Therapies

Viruses have not evolved <u>specific</u> mechanisms to evade antiviral drugs—these drugs simply haven't been around long enough to allow this type of evolution to take place. For example, HIV-1 has not evolved a protein that specifically interacts with AZT and inactivates it. However, during replication, most viruses mutate rapidly, and these mutations can render antiviral drugs ineffective. Indeed, drug-resistant mutants have been described for every antiviral drug in common use today.

It is important to remember that viral mutants arise only when viruses replicate. This means that if a drug does not stop viral replication completely, a virus will be free to mutate so as to "escape" the effects of the drug. Escape mutants are not such a big problem in acute infections (e.g., an influenza infection), because such infections are usually dealt with so quickly by host defenses that the probability of an escape mutant arising is relatively small. In contrast, with chronic infections, there is usually enough time (i.e., enough viral replication cycles) to insure that an escape mutant will eventually be produced. For example, as early as 1989, it was observed that AIDS patients who had been treated with AZT for six months or longer frequently harbored escape mutants which were resistant to this drug. The realization that a single drug would probably not be effective in controlling an HIV-1 infection led to the suggestion that multiple drugs should be used simultaneously to treat AIDS patients. The hypothesis was that if a patient were treated with several drugs at the same time, it would be less likely that a single viral genome would mutate in such a way as to simultaneously confer resistance to all the drugs. Although this hypothesis has not proved to be true in general, certain combinations of drugs have been found which can greatly decrease HIV-1 replication and extend the life of an infected individual.

Initially, various combinations of two different nucleoside analogs were tested as a multi-drug treatment for HIV-1 infections. Unfortunately, some combinations of these analogs were little or no better than a single analog used alone. The reason was that because these drugs worked in the same way, mutations which conferred resistance to one analog frequently conferred resistance to both. Eventually, through trial and observation, nucleoside analog pairs were identified that did extend the period from the beginning of treatment to escape. The reason some pairs work and others don't is still a bit mysterious. However, for certain combinations (e.g., AZT and another nucleoside analog, ddI) it has been shown that mutations which confer resistance to the first drug actually reverse the effects of another mutation that causes resistance to the second drug.

HAART

Of course, with the discovery of non-nucleoside reverse transcriptase inhibitors and HIV protease inhibitors, it was natural to try to combine these with nucleoside analogs in a drug "cocktail." Treatment with such a cocktail of anti-HIV-1 drugs is referred to as highly active antiretroviral therapy (**HAART**). If one had to guess, one might predict that the best combination of three drugs to use in HAART would be one nucleoside analog, one non-nucleoside reverse transcriptase inhibitor, and one protease inhibitor. However, so far it has turned out that the best combinations are two nucleoside analogs plus either one non-nucleoside reverse transcriptase inhibitor or one protease inhibitor. Effective combinations are difficult to predict, because they depend on the exact nature of the mutations required to resist the effects of each drug.

Although HAART can extend the life of many AIDS patients and can delay the onset of opportunistic infections, it is not without its problems. First, HAART is not a cure. HIV-1 can persist for very long periods with its provirus integrated into the genomes of infected cells. Because the drugs used in HAART target virus replication, nonreplicating proviruses cannot be touched by any of the available drugs. Indeed, latently infected cells represent a reservoir from which the virus can "bounce back" once HAART is discontinued.

Second, because the drugs currently used for HAART have a short half-life in the body, dozens of pills must be taken on a regular schedule each day. If this schedule is not rigorously adhered to, the amount of viral replication can increase dramatically, and the resulting mutants can render HAART ineffective. In addition, none of these drugs is without considerable side effects, and the tendency is for patients to discontinue HAART or to adhere less strictly to the treatment schedule once they begin to feel better—and non-adherence can lead to the resumption of viral replication and eventually to drug resistance. Further, these drugs are not cheap. A year of HAART typically costs about $12,000, putting this treatment beyond the reach of many patients. Finally, the availability of a treatment for HIV-1 infections has made this virus appear less frightening to some, and risky behavior has begun to increase among North American gay men. In fact, HAART, especially when not strictly followed, may serve to increase the length of time during which an infected individual can spread the virus to others.

Interrupted HAART

One of the negative effects of HAART is that, although the number of $CD4^+$ helper T cells increases during treatment, the number of HIV-1-specific killer T cells decreases dramatically. This is presumably because killer T cells must encounter cells that are producing new viruses in order to remain activated. And, of course, the whole point of HAART is to reduce viral replication. This decrease in the number of activated killer T cells can be a real problem, because when mutants eventually arise that can resist HAART, there will be very few killer T cells available to deal with the cells which produce these escape mutants.

Clearly the ideal situation would be one in which HAART acts to complement host defenses rather than to decrease the number of virus-specific killer T cells. In an attempt to make this happen, physicians have begun to test whether scheduled interruptions of HAART can allow killer T cells to recover, while at the same time maintaining adequate helper T cell counts and low viral loads. In several small trials, this approach appears to have some merit, although fine tuning will be required to discover how long and how frequent the interruptions should be. In this same vein, trials are also underway in which HIV-1 vaccines of various types are being administered during interruptions in HAART. The idea here is to use the vaccine to strengthen the patient's immune defenses while using HAART to keep viral loads low. Only time will tell whether these strategies will be useful in producing synergy between antiviral drugs and host defenses.

Glossary

Abortive infection – An infection in which some virus proteins are made, but no infectious virus is produced.

Acute infection – An infection in which the virus is banished after a short stay in the host.

Adaptive immune system – The part of the immune system that adapts to invaders by selecting for proliferation those B and T cells which have receptors that recognize the invader.

Antigenic drift – Small genetic changes due to errors made in copying the viral genome.

Antigenic shift – Drastic genetic changes introduced when gene segments derived from humans, birds, and animals are mixed and matched.

Capsid – A shell-like structure, composed of many copies of a few viral proteins.

Chronic infection – A long-term infection in which new viruses are produced, and the host defenses battle to keep the infection in check.

Cytolytic virus – A virus that kills the cells it infects.

Endemic virus – A virus that has established a successful virus-host relationship in a particular environment.

Endothelial cells – Cells that line the insides of blood vessels.

Enteric virus – A virus that infects the digestive tract.

Epithelial cells – Cells that line the surfaces of the body, including the skin, the respiratory tract, the digestive tract, and the reproductive tract.

Genome – The sum total of all the genetic information contained in an organism. For example, a virus' genome is the collection of all the virus' genes.

Host range – The type of cell or species a virus infects.

Innate immune system – The "hard-wired" immune system that all animals have. This defense is focused on a small number of common invaders.

Latent infection – An infection in which the virus lies dormant.

Opsonized – "Decorated" with fragments of complement proteins or antibodies.

Productive infection – An infection in which viral reproduction takes place and infectious viruses are produced.

Proliferation – The process in which cells grow to twice their original size, duplicate their genomes, and divide to produce two daughter cells.

Protease – An enzyme that cleaves proteins.

Provirus – A cDNA copy of a retroviral genome (or the DNA of a DNA virus) that is pasted into one of the chromosomes of an infected cell.

Resting Cells – Cells that are not proliferating.

Secrete – To export out of a cell into the surroundings.

Tropism – Affinity of a virus for a particular cell type.

Virion – Synonymous with virus.

Virulence – How effective a virus is in causing disease—synonymous with pathogenicity.

Virus particle – Synonymous with virus.

Index

Page numbers followed by t indicate tables.